AF556396

ENCYCLOPAEDIA OF INDUSTRIAL CHEMISTRY

ENCYCLOPAEDIA OF
INDUSTRIAL CHEMISTRY
Vol. 1

Shahzad Ahmad

ANMOL PUBLICATIONS PVT. LTD.
NEW DELHI - 110 002 (INDIA)

ANMOL PUBLICATIONS PVT. LTD.
H.O.: 4374/4B, Ansari Road, Darya Ganj,
New Delhi-110 002 (India)
Ph.: 23278000, 23261597
B.O.: No. 1015, Ist Main Road, BSK IIIrd Stage
IIIrd Phase, IIIrd Block,
Bangalore - 560 085 (India)
Visit us at: www.anmolpublications.com

Encyclopaedia of Industrial Chemistry

First Published, 2008
ISBN 978-81-261-3531-8 (Set)

PRINTED IN INDIA

Printed at Mehra Offset Press, Delhi.

Contents

Preface

Industrial chemistry is a branch of chemistry. In industrial chemistry, we study about compounds or elements, their properties, and their applications; which are used in industries. Since the time of Industrial Revolution, human intellect throughout the civilized world has been driving this Chemical Revolution. Chemical Revolution has been a continuing series of interconnected developments, growing like a string of firecrackers, which show no sign of letting up. Chemistry has succeeded almost too well. Some of its products have become so commonplace and necessary that they are inevitable for the human beings, while, some others are viewed with contempt by many. On the other hand, accidents and poor waste management, together with sensationalised media coverage, have made the industry highly visible and have sparkled public threat.

The chemical industry is comprised of the factories, which produce industrial chemicals. Till today, approximately there are 70,000 chemical products. Polymers and plastics, especially polyethylene, polystyrene and polycarbonate comprise about 80 per cent of the industries' output worldwide. Chemicals are used to make wide varieties of consumer goods, as well as thousands of items for agriculture, manufacturing, construction and service industries. The chemical industry itself consumes 26 per cent of its own output. Major industrial produces include rubber and plastic products, textiles, apparel, petroleum, pulp and paper, and primary metals.

In Europe, especially Germany, the chemical, plastics and rubber sectors are among the largest industrial sectors. The chemical industry has shown rapid growth for more than fifty years. The fastest growing areas have been in the manufacture of synthetic organic polymers used as plastics, fibres and elastomers. Chemical industry has been centralised mainly in three areas of the world, Western Europe, North America and Japan. Jointly these areas are known as "the Triad." Industrial processes are the key components of heavy industry.

This exclusive study has a complete account of all aspects of the industrial chemistry in an objective manner. Hopefully, these volumes would be able to fill the room for an exhaustive work, to prove to be beneficial for students scholars and general readers.

Editor

1

Introduction

Chemical Revolution has been a continuing series of interconnected developments, going off like a string of Chinese firecrackers, which show no signs of letting up. If steam leveraged the Industrial Revolution, human intellect throughout the civilized world has been driving this Chemical Revolution.

At the beginning of this period, chemistry, if it could be called that, was not all that far removed from alchemy. And there was no such thing as a "chemical industry." People did things with things that seemed to work: They made soap from animal fat and plant ashes; they derived substances with apparent therapeutic value from bark, roots, or plants; they made glass from sand, soda ash, and other materials; they made do with mineral colorants (iron oxide, for example) much as did their Stone Age ancestors; they used natural preservatives and fertilizers. But they did not understand why or, for the most part, how these substances worked. And they were restricted to mixing the basic materials at hand, in the basic form in which they existed at hand. That all started to change a little over two hundred years ago, when humans began to unravel the basic mysteries of matter—how it is subdivided, how it is held together, how it is transformed, how it can be reformed into "new" substances.

Unfortunately, much of the written history of chemistry, chemical technology, and the chemical industry has been fragmented and piecemeal. This is a pity, inasmuch as chemistry, more than any other physical science, has touched people directly and beneficially in almost every aspect of their daily lives through the vast array of products and systems produced by the chemical industry.

Chemistry has succeeded almost too well. Some of its products have become so commonplace and necessary that they are "invisible" to the public or even viewed with contempt by some. On the other hand, accidents and poor waste management, together with sensationalised media coverage, have made the industry highly visible and have sparked public fear. Much misinformation about chemicals and the chemical industry has been issued in recent years. Many people have been understandably unsettled.

The chemical industry comprises the companies that produce industrial chemicals. It is central to modern world economy, converting raw materials (oil, natural gas, air, water, metals, minerals) into more than 70,000 different products. Polymers and plastics, especially polyethylene, polypropylene, polyvinyl chloride, polyethylene terephthalate, polystyrene and polycarbonate comprise about 80 per cent of the industry's output worldwide.

Chemicals are used to make a wide variety of consumer goods, as well as thousands inputs to agriculture, manufacturing, construction, and service industries. The chemical industry itself consumes 26 per cent of its own output. Major industrial customers include rubber and plastic products, textiles, apparel, petroleum refining, pulp and paper, and primary metals. Chemicals is nearly a $2 trillion global enterprise, and the EU and US chemical companies are the world's largest producers. The largest corporate producers worldwide, with plants in numerous countries, are BASF, Dow, Shell, Bayer, INEOS, ExxonMobil, DuPont, and Mitsubishi, along with thousands of smaller firms.

In the US there are 170 major chemical companies. They operate internationally with more than 2,800 facilities outside the US and 1,700 foreign subsidiaries or affiliates operating. The US chemical output is $400 billion a year.

The US industry records large trade surpluses and employs more than a million people in the United States alone. The chemical industry is also the second largest consumer of energy in manufacturing and spends over $5 billion annually on pollution abatement.

In Europe, especially Germany, the chemical, plastics and rubber sectors are among the largest industrial sectors. Together they generate about 3.2 million jobs in more than 60,000 companies. Since 2000 the chemical sector alone has represented 2/3 of the entire manufacturing trade surplus of the EU. The chemical sector accounts for 12 per cent of the EU manufacturing industry's added value.

The chemical industry has shown rapid growth for more than fifty years. The fastest growing areas have been in the manufacture of synthetic organic polymers used as plastics, fibres and elastomers. Historically and presently the chemical industry has been concentrated in three areas of the world, Western Europe, North America and Japan (the Triad). The European Community remains the largest producer area followed by the USA and Japan.

The traditional dominance of chemical production by the Triad countries is being challenged by changes in feedstock availability and price, labour cost, energy cost, differential rates of economic growth and environmental pressures. Instrumental in the changing structure of the global chemical industry has been the growth in China, India, Korea, the Middle East, South East Asia, Nigeria, Trinidad, Thailand, Brazil, Venezuela, and Indonesia.

Technology

As accepted by chemical engineers, the chemical industry involves the use of chemical processes such as chemical reactions and refining methods to produce a wide variety of solid, liquid, and gaseous materials. Most of these products are used in manufacture of other items, although a smaller number are used directly by consumers. Solvents, pesticides, lye, washing soda, and portland cement are a few examples of product used by consumers.

The industry includes manufacturers of inorganic- and organic-industrial chemicals, ceramic products, petrochemicals, agrochemicals, polymers and rubber (elastomers), oleochemicals (oils, fats, and waxes), explosives, fragrances and flavours. Examples of these products are shown in the Table below:

Product Type	*Examples*
Inorganic industrial	Ammonia, nitrogen, sodium hydroxide, sulphuric acid
Organic industrial	Acrylonitrile, phenol, ethylene oxide, urea
Ceramic products	Silica brick, frit
Petrochemicals	Benzene, ethylene, styrene
Agrochemicals	Fertilizers, insecticides, herbicides
Polymers	Polyethylene, Bakelite, polyester
Elastomers	Polyisoprene, neoprene, polyurethane
Oleochemicals	Lard, soybean oil, stearic acid
Explosives	Nitroglycerin, ammonium nitrate, nitrocellulose
Fragrances and flavours	Benzyl benzoate, coumarin, vanillin

Although the pharmaceutical industry is often considered a chemical industry, it has many different characteristics that puts it in a separate category. Other closely related industries include petroleum, glass, paint, ink, sealant, adhesive, and food processing manufacturers.

Chemical processes such as chemical reactions are used in chemical plants to form new substances in various types of reaction vessels. In many cases the reactions are conducted in special corrosion resistant equipment at elevated temperatures and pressures with the use of catalysts. The products of these reactions are separated using a variety of techniques including distillation especially fractional distillation, precipitation, crystallisation, adsorption, filtration, sublimation, and drying. The processes and product are usually tested during and after manufacture by dedicated instruments and on-site quality control laboratories to insure safe operation and to assure that the product will meet required specifications.

The products are packaged and delivered by many methods, including pipelines, tank-cars, and tank-trucks (for both solids and liquids), cylinders, drums, bottles, and boxes. Chemical companies often have a research and development laboratory for developing and testing products and processes. These facilities may include pilot plants, and such research facilities may be located at a site separate from the production plant.

History

Chandler (2005) argues the relative success or failure of American and European chemical companies is explained with reference to three themes: "barriers to entry," "strategic boundaries," and "limits to growth." He says successful chemical firms followed definite "paths of learning" whereby first movers and close followers created entry barriers to would-be rivals by building "integrated learning bases" (or organisational capabilities) which enabled them to develop, produce, distribute, and sell in local and then worldwide markets. Also they followed a "virtuous strategy" of reinvestment of retained earnings and growth through diversification, particularly to utilise "dynamic" scale and scope economies relating to new learning in launching "next generation" products.

Companies in 21st Century

The chemical industry includes large, medium, and small companies that are located worldwide. For some of these companies the chemical sales represented only a portion of their total sales; for example ExxonMobil's chemical sales were only 8.7 per cent of their total sales.

Biochemical Engineering

Biochemical engineering is a branch of chemical engineering that mainly deals with the design and construction of unit processes that involve biological organisms or molecules. Biochemical engineering is often taught as a supplementary option to chemical engineering due to the similarities in both the background subject curriculum and problem-solving techniques used by both professions. Its applications are used in the pharmaceutical, biotechnology, and water treatment industries.

The Bioreactor

A bioreactor may refer to any device or system that supports a biologically active environment. In one case, a bioreactor is a vessel in which is carried out a chemical process which involves organisms or biochemically active substances derived from such organisms. This process can either be aerobic or anaerobic. These bioreactors are commonly cylindrical, ranging in size from some litre to cube metres, and are often made of stainless steel.

A bioreactor may also refer to a device or system meant to grow cells or tissues in the context of cell culture. These devices are being developed for use in tissue engineering.

On the basis of mode of operation, a bioreactor may be classified as batch, fed batch or continuous (e.g. Continuous stirred-tank reactor model). An example of a bioreactor is the chemostat.

Organisms growing in bioreactors may be suspended or immobilised. The simplest, where cells are immobilised, is a petri dish with agar gel. Large scale immobilised cell bioreactors are:

- Packed bed
- Fibrous bed
- Membrane

Bioreactor Design

Bioreactor design is quite a complex engineering task. Under optimum conditions the microorganisms or cells are able to perform their desired function with great efficiency. The bioreactor's environmental conditions like gas (i.e., air, oxygen, nitrogen, carbon dioxide) flowrates, temperature, pH and dissolved oxygen levels, and agitation speed/circulation rate need to be closely monitored and controlled.

Most industrial bioreactor manufacturers use vessels, sensors, controllers, and a control system, networked together for their bioreactor system.

Fouling can harm the overall sterility and efficiency of the bioreactor, especially the heat exchangers. To avoid it the bioreactor must be easily cleanable and must be as smooth as possible (therefore the round shape).

A heat exchanger is needed to maintain the bioprocess at a constant temperature. Biological fermentation is a major source of heat, therefore in most cases bioreactors need water refrigeration. They can be refrigerated with an external jacket or, for very large vessels, with internal coils.

In an aerobic process, optimal oxygen transfer is perhaps the most difficult task to accomplish. Oxygen is poorly soluble in water—and even less in fermentation broths—and is relatively scarce in air (20.8 per cent). Oxygen transfer is usually helped by agitation, that is also needed to mix nutrients and to keep the fermentation homogeneous. There are however limits to the speed of agitation, due to both high power consumption and the damage to organisms due to excessive tip speed.

Industrial bioreactors usually employ bacteria or other simple organisms that can withstand the forces of agitation. They are also simple to sustain, requiring only simple nutrient solutions and can grow at astounding rates.

In bioreactors where the goal is grow cells or tissues for experimental or therapeutic purposes, the design is significantly different from industrial bioreactors. Many cells and tissues, especially mammalian, must have a surface or other structural support in order to grow, and agitated environments are often destructive to these cell types and tissues. Higher organisms also need more complex growth medium.

NASA Tissue Cloning Bioreactor

NASA has developed a new type of bioreactor that artificially grows tissue in cell cultures. NASA's tissue bioreactor can grow heart tissue, skeletal tissue, ligaments, cancer tissue for study, and other types of tissue.

Biomedical Engineering

Biomedical engineering (BME) is the application of engineering principles and techniques to the medical field. It combines the design and problem solving expertise of engineering with the medical expertise of physicians to help improve patient health care and the quality of life of healthy individuals.

As a relatively new discipline, much of the work in biomedical engineering consists of research and development, covering an array of fields: bioinformatics, medical imaging, image processing, physiological signal processing, biomechanics, biomaterials and bioengineering, systems analysis, 3-D modelling, etc.

Examples of concrete applications of biomedical engineering are the development and manufacture of biocompatible prostheses, medical devices, diagnostic devices and imaging equipment such as MRIs and EEGs, and pharmaceutical drugs.

Disciplines in Biomedical Engineering

Biomedical engineering is widely considered an interdisciplinary field, resulting in a broad spectrum of disciplines that draw influence from various fields and sources. Due to the extreme diversity, it is not atypical for a biomedical engineer to focus on a particular aspect. There are many different taxonomic breakdowns of BME, one such listing defines the aspects of the field as such:

- Bioelectrical and neural engineering
- Biomedical imaging and biomedical optics
- Biomaterials
- Biomechanics and biotransport
- Biomedical devices and instrumentation
- Molecular, cellular and tissue engineering
- Systems and integrative engineering

In other cases, disciplines within BME are broken down based on the closest association to another, more established engineering field, which typically include:

- *Chemical Engineering:* Often associated with biochemical, cellular, molecular and tissue engineering, biomaterials, and biotransport.
- *Electrical Engineering:* Often associated with bioelectrical and neural engineering, bioinstrumentation, biomedical imaging, and medical devices.
- *Mechanical Engineering:* Often associated with biomechanics, biotransport, medical devices, and modelling of biological systems.
- *Optics and Optical Engineering:* Biomedical optics, imaging and medical devices.

Clinical Engineering

Clinical engineering is a branch of biomedical engineering for professionals responsible for the management of medical equipment in a hospital. The tasks of a clinical engineer are typically the acquisition and management of medical device inventory, supervising biomedical engineering technicians (BMETs), ensuring that safety and regulatory issues are taken into consideration and serving as a technological consultant for any issues in a hospital where medical devices are concerned. Clinical engineers work closely with the IT department and medical physicists.

A typical biomedical engineering department does the corrective and preventive maintenance on the medical devices used by the hospital, except for those covered by a warranty or maintenance agreement with an external company. All newly acquired equipment is also fully tested. That is, every line of software is executed, or every possible setting is exercised and verified. Most devices are intentionally simplified in some way to make the testing process less expensive, yet accurate. Many biomedical devices need to be sterilised. This creates a unique set of problems, since most sterilisation techniques can cause damage to machinery and materials. Most medical devices are either inherently safe, or have added devices and systems so that they can sense their failure and shut down into an unusable, thus very safe state. A typical, basic requirement is that no single failure should cause the therapy to become unsafe at any point during its life-cycle.

Medical Devices

A medical device is intended for use in:

- The diagnosis of disease or other conditions, or
- In the cure, mitigation, treatment, or prevention of disease,
- Intended to affect the structure or any function of the body of man or other animals, and which does not achieve any of its primary intended purposes through chemical action and which is not dependent upon being metabolised for the achievement of any of its primary intended purposes.

Some examples include pacemakers, infusion pumps, the heart-lung machine, dialysis machines, artificial organs, implants, artificial limbs, corrective lenses, cochlear implants, ocular prosthetics, facial prosthetics, somato prosthetics, and dental implants.

Stereolithography is a practical example on how medical modelling can be used to create physical objects. Beyond modelling organs and the human body, emerging engineering techniques are also currently used in the research and development of new devices for innovative therapies, treatments, patient monitoring, and early diagnosis of complex diseases.

Medical devices can be regulated and classified (in the US) as shown below:

1. Class I devices present minimal potential for harm to the user and are often simpler in design than Class II or Class III devices. Devices in this category include tongue depressors, bedpans, elastic bandages, examination gloves, and hand-held surgical instruments and other similar types of common equipment.
2. Class II devices are subject to special controls in addition to the general controls of Class I devices. Special controls may include special labelling requirements, mandatory performance standards, and postmarket surveillance. Devices in this class are typically non-invasive and include x-ray machines, PACS, powered wheelchairs, infusion pumps, and surgical drapes.
3. Class III devices require pre-market approval, a scientific review to ensure the device's safety and effectiveness, in addition to the general controls of Class I. Examples include replacement heart valves, silicone gel-filled breast implants, implanted cerebellar stimulators, implantable pacemaker pulse generators and endosseous (intra-bone) implants.

Medical Imaging

Imaging technologies are often essential to medical diagnosis, and are typically the most complex equipment found in a hospital including:

- Fluoroscopy
- Magnetic resonance imaging (MRI)
- Nuclear Medicine
- Positron Emission Tomography (PET) PET scans PET-CT scans

- Projection Radiography such as X-rays and CT scans
- Tomography
- Ultrasound
- Electron Microscopy

Tissue Engineering

One of the goals of tissue engineering is to create artificial organs for patients that need organ transplants. Biomedical engineers are currently researching methods of creating such organs. In one case bladders have been grown in lab and transplanted successfully into patients. Bioartificial organs, which utilise both synthetic and biological components, are also a focus area in research, such as with hepatic assist devices that utilise liver cells within an artificial bioreactor construct.

Regulatory Issues

Regulatory issues are never far from the mind of a biomedical engineer. To satisfy safety regulations, most biomedical systems must have documentation to show that they were managed, designed, built, tested, delivered, and used according to a planned, approved process. This is thought to increase the quality and safety of diagnostics and therapies by reducing the likelihood that needed steps can be accidentally omitted again.

In the United States, biomedical engineers may operate under two different regulatory frameworks. Clinical devices and technologies are generally governed by the Food and Drug Administration (FDA) in a similar fashion to pharmaceuticals. Biomedical engineers may also develop devices and technologies for consumer use, such as physical therapy devices, which may be governed by the Consumer Product Safety Commission.

Other countries typically have their own mechanisms for regulation. In Europe, for example, the actual decision about whether a device is suitable is made by the prescribing doctor, and the regulations are to assure that the device operates as expected. Thus in Europe, the governments license certifying agencies, which are for-profit. Technical committees of leading engineers write recommendations which incorporate public comments and are adopted as regulations by the European Union. These recommendations vary by the type of device, and specify tests for safety and efficacy.

Once a prototype has passed the tests at a certification lab, and that model is being constructed under the control of a certified quality system, the device is entitled to bear a CE mark, indicating that the device is believed to be safe and reliable when used as directed.

The different regulatory arrangements sometimes result in technologies being developed first for either the US or in Europe depending on the more favourable form of regulation. Most safety-certification systems give equivalent results when applied diligently. Frequently, once one such system is satisfied, satisfying the other requires only paperwork.

Biomedical Engineering Training

Education: Biomedical engineers combine sound knowledge of engineering and biological science, and therefore tend to have a bachelors of science and advanced degrees from major universities, who are now improving their biomedical engineering curriculum because interest in the field is increasing. Many colleges of engineering now have a biomedical engineering programme or department from the undergraduate to the doctoral level.

Traditionally, biomedical engineering has been an interdisciplinary field to specialise in after completing an undergraduate degree in a more traditional discipline of engineering or science, the reason for this being the requirement for biomedical engineers to be equally knowledgeable in engineering and the biological sciences. However, undergraduate programmes of study combining these two fields of knowledge are becoming more widespread, including programmes for a Bachelor of Science in Biomedical Engineering. As such, many students also pursue an undergraduate degree in biomedical engineering as a foundation for a continuing education in medical school. Though the number of biomedical engineers is currently low (as of 2004, under 10,000 in the US), the number is expected to rise as modern medicine and technology improves.

In the US, an increasing number of undergraduate programmes are also becoming recognised by ABET as accredited bioengineering/biomedical engineering programmes. Over 40 programmes are currently accredited by ABET, the first being Duke University, originally accredited by the Engineering Council for Profession Development (now ABET) in September of 1972.

As with many degrees, the reputation and ranking of a programme may factor into the desirability of a degree holder for either employment or graduate admission. The reputation of many undergraduate degrees are also linked to the institution's graduate or research programmes, which have some tangible factors for rating, such as research funding and volume, publications and citations.

Graduate education is also an important aspect in BME. Although many engineering professions do not require graduate level training, BME professions often recommend or require them. Since many BME professions often involve scientific research, such as in the pharmaceutical and medical device industries, graduate education may be highly desirable as undergraduate degrees typically do not provide substantial research training and experience.

Graduate programmes in BME, like in other scientific fields, are highly varied and particular programmes may emphasise certain aspects within the field. They may also feature extensive collaborative efforts with programmes in other fields, owing again to the interdisciplinary nature of BME.

Education in BME also varies greatly around the world. The US has, by virtue of being a large country with fewer internal barriers, having an extensive biotechnology sector and dozens of major universities, has progressed a great deal in the development of BME education and training. Europe, which also has a large biotechnology sector and an impressive education system, has encountered trouble in creating uniform standards as the European community attempts to bring down some of the national barriers that exist. Recently, initiatives such as BIOMEDEA have sprung up to develop BME-related education and professional standards. Other countries, such as Australia, are recognising and moving to correct deficiencies in their BME education. Also, as high technology endeavours are usually marks of developed nations, some areas of the world are prone to slower development in education, including in BME.

Professional Certification

Engineers typically require a type of professional certification, such as satisfying certain education requirements and passing an examination to become a professional engineer. These certifications are usually nationally regulated and registered, but there are also cases where a self-governing body, such as the Canadian Association of Professional Engineers. In many cases, carrying the title of "Professional Engineer" is legally protected.

As BME is an emerging field, professional certifications are not as standard and uniform as they are for other engineering fields. For example, the Fundamentals of Engineering exam in the US does not include a biomedical engineering section, though it does cover biology. Biomedical engineers often simply possess a university degree as their qualification. However, some countries do regulate biomedical engineers, such as Australia, however registration is typically recommended, but not always a requirement.

2

Industrial Process

Industrial processes are procedures involving chemical or mechanical steps to aid in the manufacture of an item or items, usually carried out on a very large scale.

Industrial processes are the key components of heavy industry. Most processes make the production of an otherwise rare material vastly cheaper, thus changing it into a commodity; i.e. the process makes it economically feasible for society to use the material on a large scale. One of the best examples of this is the change in aluminium from prices more expensive than silver to its use in recyclable/disposable beverage containers.

An industrial process differs from a craft, workshop or laboratory process by the scale or investment required. Most of the processes are complex and require large capital investments in machinery, or a substantial amount of raw materials, in comparison to batch or craft processes. Production of a specific material may involve more than one type of process.

General Processes

These may be applied on their own, or as part of a larger process.

Liquefaction of Gases

Liquefaction of gases includes a number of processes used to convert a gas into a liquid state. The processes are used for scientific, industrial and commercial purposes. Many gases can be put into a liquid state at normal atmospheric pressure by simple cooling; a few, such as carbon dioxide, require pressurisation as well. Liquefaction is used for analysing the fundamental properties of gas molecules (intermolecular forces), for storage of gases, for example: LPG, and in refrigeration and air conditioning. There the gas is liquefied in the *condenser*, where the heat of vaporisation is released, and evaporated in the *evaporator*, where the heat of vaporisation is absorbed. Ammonia was the first such refrigerant, but it has been replaced by compounds derived from petroleum and halogens.

Liquid oxygen is provided to hospitals for conversion to gas for patients suffering from breathing problems, and liquid nitrogen is used by dermatologists and by inseminators to freeze semen. Liquefied chlorine is transported for eventual solution in water, after which it is used for water purification, sanitation of industrial waste, sewage and swimming pools, bleaching of pulp and textiles and manufacture of carbon tetrachloride, glycol and numerous other organic compounds as well as phosgene gas. It was used in warfare in World War I at Flanders and in gaseous form at Ypres, France, though the shells were filled with liquid.

Liquefaction of helium (^4He) led to a Nobel Prize for Heike Kamerlingh Onnes in 1913. At ambient pressure the boiling point of liquefied helium is 4.22°K (-268.93°C). Below 2.17°K liquid ^4He has many amazing properties, such as climbing the walls of the vessel, exhibiting zero viscosity, and offering no lift to a wing past which it flows.

Supercritical Drying

Supercritical drying is a process to remove liquid in a precisely controlled way, similar to freeze drying. It is useful in the production of MEMS and the drying of spices, and is commonly used in the production of aerogel.

As a substance crosses the boundary from liquid to gas, the size of the liquid decreases. As this happens, surface tension at the solid-liquid interface pulls against any structures that the liquid is attached to. Delicate structures, like cell walls, the dendrites in silica gel, and the tiny machinery of MEMS devices, tend to be broken apart by this surface tension as the interface moves by.

To avoid this, the sample can be brought from the liquid phase to the gas phase without crossing the liquid-gas boundary on the phase diagram; in freeze-drying, this means going around to the left (low temperature, low pressure). However, some structures are disrupted even by the solid-gas boundary. Supercritical drying, on the other hand, goes around the line to the right, on the high-temperature, high-pressure side. This route from liquid to gas does not cross any phase boundary, instead passing through the supercritical region, where the distinction between gas and liquid ceases to apply.

Fluids suitable for supercritical drying include carbon dioxide (critical point 304.25°K at 7.39 MPa or 31.1 degrees Celsius at 1072 psi) and freon (~300°K at 3.5-4 MPa or 25 to 30°C at 500-600 psi).

Nitrous oxide has similar physical behaviour to carbon dioxide, but is a powerful oxidiser in its supercritical state. Supercritical water is also a powerful oxidiser, partly because its critical point occurs at such a high temperature (374°C) and pressure (3212 psi).

In most such processes, acetone is first used to wash away all water, exploiting the complete miscibility of these two fluids. The acetone is then washed away with high pressure carbon dioxide, the industry standard now that freon is unavailable. Again, acetone and liquid carbon dioxide are completely miscible. The carbon dioxide is then heated until its pressure goes beyond the critical point, at which time the pressure can be gradually released, allowing the gas to escape and leaving a dried product.

Freeze-drying

Freeze-drying (also known as lyophilisation) is a dehydration process typically used to preserve a perishable material or make the material more convenient for transport. Freeze drying works by freezing the material and then reducing the surrounding pressure and adding enough heat to allow the frozen water in the material to sublime directly from the solid phase to gas. There are three stages in the complete freeze-drying process: Freezing, Primary Drying, and Secondary Drying.

Freezing: The freezing process consists of freezing the material. In a lab, this is often done by placing the material in a freeze-drying flask and rotating the flask in a bath, called a shell freezer, which is cooled by mechanical refrigeration, dry ice and methanol, or liquid nitrogen. On a larger-scale, freezing is usually done using a freeze-drying machine. In this step, it is important to freeze the material at a temperature below the eutectic point of the material.

Since the eutectic point occurs at the lowest temperature where the solid and liquid phase of the material can coexist, freezing the material at a temperature below this point ensures that sublimation rather than melting will occur in the following steps. Larger crystals are easier to freeze dry. To produce larger crystals the product should be frozen slowly or can be cycled up and down in temperature. This cycling process is called annealing.

Amorphous (glassy) materials do not have a eutectic point, but do have a critical temperature, below which the product must be maintained to prevent melt-back or collapse during primary and secondary drying.

Primary Drying: During the primary drying phase the pressure is lowered and enough heat is supplied to the material for the water to sublimate. The amount of heat necessary can be calculated using the sublimating molecules' latent heat of sublimation. In this initial drying phase about 98 per cent of the water in the material is sublimated. This phase may be slow, because if too much heat is added the material's structure could be altered.

In this phase, pressure is controlled through the application of partial vacuum. The vacuum speeds sublimation making it useful as a deliberate drying process. Furthermore, a cold condenser chamber and/or condenser plates provide a surface(s) for the water vapour to resolidify on. This condenser plays no role in keeping the material frozen; rather, it prevents water vapour from reaching the vacuum pump, which could degrade the pump's performance. Condenser temperatures are typically below –50°C.

Secondary Drying: The secondary drying phase aims to sublimate the water molecules that are adsorbed during the freezing process, since the mobile water molecules were sublimated in the primary drying phase. This part of the freeze-drying process is governed by the material's adsorption isotherms. In this phase, the temperature is raised even higher than in the primary drying phase to break any physicochemical interactions that have formed between the water molecules and the frozen material. Usually the pressure is also lowered in this stage to encourage sublimation. However, there are products that benefit from increased pressure as well.

After the freeze drying process is complete, the vacuum is usually broken with an inert gas, such as nitrogen, before the material is sealed.

Properties of Freeze-dried Products

If a freeze-dried substance is sealed to prevent the reabsorption of moisture, the substance may be stored at room temperature without refrigeration, and be protected against spoilage for many years. Preservation is possible because the greatly reduced water content inhibits the action of microorganisms and enzymes that would normally spoil or degrade the substance.

Freeze drying also causes less damage to the substance than other dehydration methods using higher temperatures.

Freeze drying does not usually cause shrinkage or toughening of the material being dried. In addition, flavours and smells generally remain unchanged making the process popular for preserving food. Unfortunately, water is not the only chemical capable of sublimation and the loss of other volatile compounds such as acetic acid (vinegar) and alcohols can yield undesirable results.

Freeze-dried products can be rehydrated (reconstituted) much more quickly and easily because it leaves microscopic pores. The pores are created by the ice crystals that sublimate, leaving gaps or pores in its place. This is especially important when it comes to pharmaceutical uses. Lyophilisation can also be used to increase the shelf life of some pharmaceuticals for many years.

Uses of Freeze-drying

Freeze-drying is used in many different industries, sometimes for different reasons.

Pharmaceutical and Bio-Tech: Pharmaceutical companies often use freeze-drying to increase the shelf life of products, such as vaccines and other injectables. By removing the water from the material and sealing the material in a vial, the material can be easily stored, shipped and later reconstituted to its original form for injection.

Food Industry: The process has been popularised in the forms of freeze-dried ice cream; an example of astronaut food. It is also popular and convenient for hikers because the reduced weight allows them to carry more food and reconstitute it with available water. Instant coffee is sometimes freeze-dried, despite high costs of freeze dryers. The coffee is often dried by vaporisation in a hot air flow, or by projection on hot metallic plates. Currently, the freeze drying process is used more commonly in the pharmaceutical industry.

Technological Industry: In chemical synthesis, products are often lyophilised to make them more stable, or easier to dissolve in water for subsequent use.

In bioseparations, freeze-drying can also be used as a late-stage purification procedure, because it can effectively remove solvents. Furthermore, it is capable of concentrating molecules with low molecular weights that are too small to be filtered out by a filtration membrane.

Freeze-drying is a relatively expensive process. The equipment is about three times as expensive as the equipment used for other separation processes, and the high energy demands lead to high energy costs. Furthermore, freeze-drying also has a long process time, because the addition of too much heat to the material can cause melting or structural deformations. Therefore, freeze-drying is often reserved for materials that are heat-sensitive, such as proteins, enzymes, microorganisms, and blood plasma. The low operating temperature of the process leads to minimal damage of these heat-sensitive products.

Other Uses: Recently, some taxidermists have begun using freeze drying to preserve animals.

Organisations, such as the Document Conservation Laboratory at the United States National Archives and Records Administration (NARA), have done studies on freeze drying as a recovery method of water-damaged books and documents. While recovery is possible, restoration quality depends on the material of the documents. If a document is made of a variety of materials, which have different absorption properties, expansion will occur at a non-uniform rate which could lead to physical deformations. Water can also cause mould to grow and prompt image media to become soluble causing bleeding. In these cases, freeze drying may not be an effective restoration method.

In high altitude environments, the low temperatures and pressures can sometimes produce natural mummies by a process of freeze-drying.

Freeze-drying Equipment

There are essentially three categories of freeze dryers: rotary evaporators, manifold freeze dryers, and tray freeze dryers.

Rotary freeze dryers are usually used with liquid products, such as pharmaceutical solutions and tissue extracts.

Manifold freeze dryers are usually used when drying a large amount of small containers and the product will be used in a short period of time. A manifold dryer will dry the product to less than 5 per cent moisture content. Without heat only primary drying (removal of the unbound water) can be achieved. A heater needs to be added for secondary drying, which will remove the bound water and will produce a lower moisture content.

Tray freeze dryers are more sophisticated and are used to dry a variety of materials. A tray freeze dryer is used to produce the driest product for long term storage. A tray freeze dryer allows the product to be frozen in place and performs both primary (unbound water removal) and secondary (bound water removal) freeze drying, thus producing the driest possible end-product. Tray freeze dryers can dry product in bulk or in vials. When drying in vials, the freeze dryer is supplied with a stoppering mechanism that allows a stopper to be pressed into place sealing the vial before it is exposed to the atmosphere. This is used for long term storage, such as vaccines.

Scrubber

Scrubber systems are a diverse group of air pollution control devices that can be used to remove particulates and/or gases from industrial exhaust streams. Traditionally, the term "scrubber" has referred to pollution control devices that used liquid to "scrub" unwanted pollutants from a gas stream. Recently, the term is also used to describe systems that inject a dry reagent or slurry into a dirty exhaust stream to "scrub out" acid gases. Scrubbers are one of the primary devices that control gaseous emissions, especially acid gases.

Removal and Neutralisation

The exhaust gases of combustion may at times contain substances considered harmful to the environment, and it is the job of the scrubber to either remove those substances from the exhaust gas stream, or to neutralise those substances so that they cannot do any harm once emitted into the environment as part of the exhaust gas stream.

Wet Scrubbing

A wet scrubber is used to clean air or other gases of various pollutants and dust particles. Wet scrubbing works via the contact of target compounds or particulate matter with the scrubbing solution. Solutions may simply be water (for dust) or complex solutions of reagents that specifically target certain compounds.

Removal efficiency of pollutants is improved by increasing residence time in the scrubber or by the increase of surface area of the scrubber solution by the use of a spray nozzle, packed towers or an aspirator. Wet scrubbers will often significantly increase the proportion of water in waste gases of industrial processes which can be seen in a stack plume.

Dry Scrubbing

A dry or semi-dry scrubbing system, unlike the wet scrubber, does not saturate with moisture the flue gas stream that is being treated. In some cases no moisture is added; while in other designs only the amount of moisture that can be evaporated in the flue gas without condensing is added. Therefore, dry scrubbers do not have a stack steam plume or wastewater handling/disposal requirements. Dry scrubbing systems are used to remove acid gases (such as SO_2 and HCl) primarily from combustion sources.

There are a number of dry type scrubbing system designs. However, all consist of two main sections or devices: a device to introduce the acid gas sorbent material into the gas stream and a particulate matter control device to remove reaction products, excess sorbent material as well as any particulate matter already in the flue gas.

Dry scrubbing systems can be categorised as dry sorbent injectors (DSIs) or as spray dryer absorbers (SDAs). Spray dryer absorbers are also called semi-dry scrubbers or spray dryers.

Dry sorbent injection involves the addition of an alkaline material (usually hydrated lime or soda ash) into the gas stream to react with the acid gases. The sorbent can be injected directly into several different locations: the combustion process, the flue gas duct (ahead of the particulate control device), or an open reaction chamber (if one exists). The acid gases react with the alkaline sorbents to form solid salts which are removed in the particulate control device. These simple systems can achieve only limited acid gas (SO_2 and HCl) removal efficiencies. Higher collection efficiencies can be achieved by increasing the flue gas humidity (i.e., cooling using water spray). These devices have been used on medical waste incinerators and a few municipal waste combustors.

In spray dryer absorbers, the flue gases are introduced into an absorbing tower (dryer) where the gases are contacted with a finely atomised alkaline slurry. Acid gases are absorbed by the slurry mixture and react to form solid salts which are removed by the particulate control device. The heat of the flue gas is used to evaporate all the water droplets, leaving a non-saturated flue gas to exit the absorber tower. Spray dryers are capable of achieving high (80 per cent) acid gas removal efficiencies. These devices have been used on industrial and utility boilers and municipal waste combustors.

Mercury Removal

Mercury has no known beneficial uses in nature, but it is a common substance found in coal that must also be removed. Wet scrubbers are only effective for mercury removal under certain conditions. Mercury vapour in its elemental form, Hg^0, is insoluble in the scrubber slurry and not removed. Oxidised mercury, Hg^{2+}, compounds are more soluble in the scrubber slurry and can be captured. The type of coal burned as well as the presence of a selective catalytic reduction unit both affect the ratio of elemental to oxidised mercury in the flue gas and thus the degree to which the mercury is removed.

Scrubber Waste Products

One side effect of scrubbing is that the process only moves the unwanted substance from the exhaust gases into a solid paste or powder form. If there is no useful purpose for this solid waste, it must be either contained or buried to prevent environmental contamination. Limestone-based scrubbers can produce a synthetic gypsum of sufficient quality that can be used to manufacture drywall and other industrial products.

Mercury removal results in a waste product that either needs further processing to extract the raw mercury, or must be buried in a special hazardous wastes landfill that prevents the mercury from seeping out into the environment.

Bacteria Spread

Until recently, scrubbers have not been associated with health risks involving bacteria spread as a result of inadequate cleaning, unlike other devices such as cooling towers. However, a 2005 outbreak of Legionnaires' disease in Norway was proven to emanate from a scrubber, causing ten deaths and more than fifty cases of infection as it spread the bacteria through the air during a period of only two weeks. This particular system had undergone regular cleaning routines every three weeks. This case is believed to be the first documented case of a scrubber being the source of such bacteria spread.

Batch Rectifier

The simplest and most frequently used batch distillation configuration is the batch rectifier, including the alembic and pot still. The batch rectifier consists of a pot (or reboiler), rectifying column, a condenser, some means of splitting off a portion of the condensed vapour (distillate) as reflux, and one or more receivers.

The pot is filled with liquid mixture and heated. Vapour flows upwards in the rectifying column and condenses at the top. Usually, the entire condensate is initially returned to the column as reflux. This contacting of vapour and liquid considerably improves the separation. Generally, this step is named start-up. After some time, a part of the overhead condensate is withdrawn continuously as distillate and it is accumulated in the receivers, and the other part is recycled into the column as reflux.

Owing to the differing vapour pressures of the distillate, there will be a change in the overhead distillation with time, as early on in the batch distillation the components with a lower vapour pressure will be evaporated first. As the supply of the material is limited and lighter components are removed, the relative fraction of heavier components will increase as the distillation progresses.

Batch Stripper

The other simple batch distillation configuration is the batch stripper. The batch stripper consists of the same parts as the batch rectifier. However, in this case, the charge pot is located above the stripping column.

During operation (after charging the pot and starting up the system) the high boiling constituents are primarily separated from the charge mixture. The liquid in the pot is depleted in the high boiling constituents, and enriched in low boiling ones. The high boiling product is routed into the bottom product receivers. The residual low boiling product is withdrawn from the charge pot. This mode of batch distillation is very seldom applied in industrial processes.

Middle Vessel Column

A third feasible batch column configuration is the middle vessel column. The middle vessel column consists of both a rectifying and a stripping section and the charge pot is located at the middle of the column.

Feasibility Studies

Generally, the feasibility studies of batch distillation are based on analyses of the following maps:

- Residue Curve map

- Still path map
- Distillate path map
- Different column profile maps

During the feasibility studies, the following basic simplifying assumptions are made:

- An infinite number of equilibrium stages
- An infinite reflux ratio
- A negligible tray hold-up in the two column sections
- A quasi-steady state in the column
- A constant molar overflow

Bernot *et al.* used the batch distillation regions to determine the sequence of the fractions. According to Ewell and Welch, a batch distillation region gives the same fractions upon rectification of any mixture lying within it. Bernot *et al.* examined the still and distillate paths for the determination of the region boundaries under high number of stages and high reflux ratio, named maximal separation.

Pham and Doherty in pioneering work described the structure and properties of residue curve maps for ternary heterogeneous azeotropic mixtures. In their model, the possibility of the phase separation of the vapour condensed is not taken into consideration yet. The singular points of the residue curve maps determined by this method were used to assign batch distillation regions by Rodriguez-Donis *et al.* and Skouras *et al.* Modla *et al.* pointed out that this method may give misleading results for the minimal amount of entrainer. Lang and Modla the method of Pham and Doherty and suggested a new, general method for the calculation of residue curves and for the determination of batch distillation regions of heteroazeotropic distillation.

Lelkes *et al.* published a feasibility method for the separation of minimum boiling point azeotropes by continuously entrainer feeding batch distillation. This method has been applied for the use of a light entrainer in the batch rectifier and stripper by Lang *et al.* (1999) and it applied for maximum azeotropes by Lang *et al.* Modla *et al.* extended this method for batch heteroazeotropic distillation under continuous entrainer feeding.

Laboratory Setup

Fractional distillation in a laboratory makes use of common laboratory glassware, as well as some single-purpose items like a fractionating column and "pigs" and "cows" used to cut fractions.

Apparatus

- Heat source, such as a hot plate with a bath.
- Distilling flask, typically a round-bottom flask.
- Receiving flask, often also a round-bottom flask.
- Fractionating column (Vigreux column).
- Distillation head.
- Thermometer and adapter if needed.
- Condenser
  - — Liebig condenser,
  - — Graham condenser, or
  - — Allihn condenser.
- Vacuum adapter (not used in upper right image).
- Boiling chips, also known as anti-bumping granules.
- Rubber bungs, unless laboratory glassware with ground glass joints is used, such as quickfit apparatus.

Method

As an example, consider the distillation of a mixture of water and ethanol. Ethanol boils at 78.5°C while water boils at 100°C. So, by gently heating the mixture, the most volatile component will concentrate to a greater degree in the vapour leaving the liquid. Some mixtures form azeotropes, where the mixture boils at a lower temperature than either component. In this example, a mixture of 95 per cent ethanol and 5 per cent water boils at 78.2°C, being more volatile than pure ethanol. For this reason, ethanol cannot be completely purified by direct fractional distillation of ethanol-water mixtures.

The mixture is put into the round bottomed flask along with a few anti-bumping granules, and the fractionating column is fitted into the top. As the mixture boils, vapour rises up the column. The vapour condenses on the glass platforms, known as trays, inside the column, and runs back down into the liquid below, refluxing distillate. The column is heated from the bottom. The hottest tray is at the bottom and the coolest is at the top. At steady state conditions, the vapour and liquid on each tray are at *equilibrium*.

Only the most volatile of the vapours stays in gaseous form all the way to the top. The vapour at the top of the column, then passes into the condenser, which cools it down until it liquefies. The separation is more pure with the addition of more trays (to a practical limitation of heat, flow, etc.) The condensate that was initially very close to the azeotrope composition becomes gradually richer in water. The process continues until all the ethanol boils out of the mixture. This point can be recognised by the sharp rise in temperature shown on the thermometer.

In laboratory distillation, several types of condensers are commonly found. The Liebig condenser is simply a straight tube within a water jacket, and is the simplest (and relatively least expensive) form of condenser. The Graham condenser is a spiral tube within a water jacket, and the Allihn condenser has a series of large and small constrictions on the inside tube, each increasing the surface area upon which the vapour constituents may condense.

Alternate setups may utilise a "cow" or "pig" which is connected to three or four receiving flasks. By turning the "cow" or "pig", the distillates can be channelled into the appropriate receiver.

Vacuum distillation systems operate at reduced pressure, thereby lowering the boiling point of the materials.

Industrial Distillation

Distillation is the most common form of separation technology used in petroleum refineries, petrochemical and chemical plants and natural gas processing plants. In most cases, the distillation is operated at a continuous steady state. New feed is always being added to the distillation column and products are always being removed. Unless the process is disturbed due to changes in feed, heat, ambient temperature, or condensing, the amount of feed being added and the amount of product being removed are normally equal. This is known as continuous, steady-state fractional distillation.

Industrial distillation is typically performed in large, vertical cylindrical columns known as "distillation or fractionation towers" or "distillation columns" with diameters ranging from about 65 centimetres to 6 metres and heights ranging from about 6 metres to 60 metres or more.

The distillation towers have liquid outlets at intervals up the column which allow for the withdrawal of different *fractions* or products having different boiling points or boiling ranges. The "lightest" products (those with the lowest boiling point) exit from the top of the columns and the "heaviest" products (those with the highest boiling point) exit from the bottom of the column.

For example, fractional distillation is used in oil refineries to separate crude oil into useful substances (or fractions) having different hydrocarbons of different boiling points. The crude oil fractions with higher boiling points:

- Have more carbon atoms.
- Have higher molecular weights.
- Are darker in colour.
- Are more viscous.
- Are more difficult to ignite and to burn.

Large-scale industrial towers use reflux to achieve a more complete separation of products. Reflux refers to the portion of the condensed overhead liquid product from a distillation or fractionation tower that is returned to the upper part of the tower as shown in the schematic diagram of a typical, large-scale industrial distillation tower. Inside the tower, the reflux liquid flowing downwards provides the cooling needed to condense the vapours flowing upwards, thereby increasing the effectiveness of the distillation tower. The more reflux is provided for a given number of theoretical plates, the better the tower's separation of lower boiling materials from higher boiling materials. Alternatively, the more reflux provided for a given desired separation, the fewer theoretical plates are required.

Fractional distillation is also used in air separation, producing liquid oxygen, liquid nitrogen, and high purity argon. Distillation of chlorosilanes also enable the production of high-purity silicon for use as a semi-conductor.

In industrial uses, sometimes a packing material is used in the column instead of trays, especially when low pressure drops across the column are required, as when operating under vacuum. This packing material can either be random dumped packing (1-3" wide) such as Raschig rings or structured sheet metal. Typical manufacturers are Koch, Sulzer and other companies. Liquids tend to wet the surface of the packing and the vapours pass across this wetted surface, where mass transfer takes place. Unlike conventional tray distillation in which every tray represents a separate point of vapour liquid equilibrium he vapour liquid equilibrium curve in a packed column is continuous. However, when modelling packed columns it is useful to compute a number of "theoretical plates" to denote the separation efficiency of the packed column with respect to more traditional trays. Differently shaped packings have different surface areas and void space between packings. Both of these factors affect packing performance.

Design of Industrial Distillation Columns

Design and operation of a distillation column depends on the feed and desired products. Given a simple, binary component feed, analytical methods such as the McCabe-Thiele method or the Fenske equation can be used to assist in the design.

Moreover, the efficiencies of the vapour-liquid contact devices (referred to as *plates* or *trays*) used in distillation columns are typically lower than that of a theoretical 100 per cent efficient equilibrium stage. Hence, a distillation column needs more plates than the number of theoretical vapour-liquid equilibrium stages.

An indication of numbers: the separation of two compounds with relative volatility of 1.1 requires a minimum of 130 theoretical plates with a minimum reflux ratio of 20. With a relative volatility of 4, the required number of theoretical plates decreased to 9 with a reflux ratio of 0.66. In another source, a boiling point difference of 30°C requires 12 theoretical plates and, for a difference of 3°C, the number of plates increased to 1000.

The reflux ratio is the ratio of the amount of moles returned as refluxed liquid to the fractionating column and the amount of moles of final product, both per unit time.

Steam Distillation

Steam distillation is a special type of distillation (a separation process) for temperature sensitive materials like natural aromatic compounds.

Many organic compounds tend to decompose at high sustained temperatures. Separation by normal distillation would then not be an option, so water or steam is introduced into the distillation apparatus. By adding water or steam the boiling points of the compounds are depressed, allowing them to evaporate at lower temperatures, preferably below the temperatures at which the deterioration of the material becomes appreciable. If the substances to be distilled are very sensitive to heat, steam distillation can also be combined with vacuum distillation. After distillation the vapours are condensed as usual, usually yielding a two-phase system of water and the organic compounds, allowing for simple separation.

Steam distillation is employed in the manufacture of essential oil, for instance, perfumes. In this method steam is passed through the plant material containing the desired oils. It is also employed in the synthetic procedures of complex organic compounds. Eucalyptus oil and orange oil are obtained by this method in industrial scale.

Steam distillation is also widely used in petroleum refineries and petrochemical plants where it is commonly referred to as "steam stripping".

Vacuum Distillation

Vacuum distillation is a method of distillation whereby the pressure above the liquid mixture to be distilled is reduced to less than its vapour pressure (usually less than atmospheric) causing evaporation of the least volatile liquid(s) (those with the highest boiling points). Vacuum distillation is used with or without heating the solution.

Vacuum distillation works on the principle that boiling occurs when the vapour pressure of a liquid exceeds the ambient pressure. Ambient pressure refers to the atmospheric pressure for an open system, or to the pressure in the distillation apparatus, for a closed system.

The process is used when liquids to be distilled have high atmospheric boiling points or chemically change at temperatures near their atmospheric boiling points. Temperature sensitive materials (such as beta carotene) also require vacuum distillation to remove solvents from the mixture without damaging the product. Another reason vacuum distillation is used is that compared to atmospheric distillation there is a lower level of residue build up. This is important in commercial applications where temperature transfer is produced using heat exchangers.

Vacuum distillation is sometimes referred to as low temperature distillation. Typical industrial applications utilise the heat pump cycle to maximise efficiency. Common standards for oil industry are ASTM D1160, D2892, D5236.

These standards describe typical applications of vacuum distillation at pressures ~1-100 mbar. Pilot plants up to 200 L can be built in accordance with these standards.

Airsensitive Vacuum Distillation

Some compounds have high boiling points as well as being airsensitive. A simple vacuum distillation system as exemplified above can be used, whereby the vacuum is replaced with an inert gas after the distillation is complete. However, this is a less satisfactory system if one desires to collect fractions under a reduced pressure. To do this a "pig" adaptor can be added to the end of the condenser, or for better results or for very air sensitive compounds a Perkin triangle apparatus can be used.

The Perkin triangle, has means via a series of glass or teflon taps to allows fractions to be isolated from the rest of the still, without the main body of the distillation being removed from either the vacuum or heat source, and thus can remain in a state of reflux. To do this, the sample is first isolated from the vacuum by means of the taps, the vacuum over the sample is then replaced with an inert gas (such as nitrogen or argon) and can then be stoppered and removed. A fresh collection vessel can then be added to the system, evacuated and linked back into the distillation system via the taps to collect a second fraction, and so on, until all fractions have been collected.

Azeotropic Distillation

In chemistry, azeotropic distillation is any of a range of techniques used to break an azeotrope in distillation. In chemical engineering, azeotropic distillation usually refers to the specific technique of adding another component to generate a new lower-boiling azeotrope that is heterogeneous, such as the example below with the addition of benzene to water and ethanol.

Example: Distillation of Ethanol/water

A common distillation with an azeotrope is the distillation of ethanol and water. Using normal distillation techniques, ethanol can only be purified to approximately 96 per cent (hence the 96 per cent (192 proof) strength of some commercially available grain alcohols).

Once at a 96/4 per cent ethanol/water concentration the vapour from the boiling mixture is also 96/4 per cent. Further distillation is therefore ineffective. Some uses require a higher percentage of alcohol, for example when used as a gasoline additive. The 96/4 per cent azeotrope needs to be "broken" in order to refine further.

Material Separation Agent

One method is the addition of an "MSA", a material separation agent. The addition of benzene to the mixture changes the molecular interactions and eliminates ("breaks") the azeotrope. The drawback is that another separation is needed to remove the benzene.

Pressure-swing Distillation

Another method, pressure-swing distillation, relies on the fact that an azeotrope is pressure dependent. It also depends on the knowledge that an azeotrope is not a range of concentrations that can not be distilled, but the point at which activity coefficients are crossing one another. If the azeotrope can be "jumped over", distillation can continue, although because the activity coefficients have crossed, the water will boil out of the ethanol.

To "jump" the azeotrope, the azeotrope can be moved by altering the pressure. Typically, pressure will be set such that the azeotrope will be closer to 100 per cent concentration. For ethanol, that may be 97 per cent. Ethanol can now be distilled up to 97 per cent. It will actually be distilled to something slightly less, like 96.5 per cent. The 96.5 per cent alcohol is then sent to a distillation column that is under a different pressure, one that pulls the azeotrope down, maybe to 96 per cent. Since the mixture is already above the 96 per cent azeotrope, the distillation will not get "stuck" at that point and the ethanol can be distilled to whatever concentration is needed.

Molecular Sieves

For the distillation of ethanol for gasoline addition, the most common means of breaking the azeotrope is the use of molecular sieves. Ethanol is distilled to 96 per cent, then run over a molecular sieve which absorbs water from the mixture. The concentration is now above 96 per cent and can be further distilled. The sieve is heated to remove the water and reused.

Short Path Distillation

Short path distillation is a distillation technique that involves the distillate travelling a short distance, often only a few centimetres. A classic example would be a distillation involving the distillate travelling from one glass bulb to another, without the need for a condenser separating the two chambers. This technique is often used for compounds which are unstable at high temperatures. Advantages are that the temperature of the boiling liquid does not have to be much higher than the boiling point of the distilling substance, and the gases only have to travel a short distance while in the gas-phase before they can be cooled again to a lower temperature.

Other Types

- In rotary evaporation a vacuum distillation apparatus is used to remove bulk solvents from a sample. Typically the vacuum is generated by a water aspirator or a membrane pump.
- In a kugelrohr a short path distillation apparatus is typically used (generally in combination with a (high) vacuum) to distil high boiling (> 300°C) compounds. The apparatus consists of an oven in which the compound to be distilled is placed, a receiving portion which is outside of the oven, and a means of rotating the sample. The vacuum is normally generated by using a high vacuum pump.
- The process of reactive distillation involves using the reaction vessel as the still. In this process, the product is usually significantly lower-boiling than its reactants. As the product is formed from the reactants, it is

vaporised and removed from the reaction mixture. This technique is an example of a continuous vs. a batch process; advantages include less downtime to charge the reaction vessel with starting material, and less workup.

- Destructive distillation involves the strong heating of solids (often organic material) in the absence of oxygen (to prevent combustion) to evaporate various high-boiling liquids, as well as thermolysis products. The gases evolved are cooled and condensed as in normal distillation. The destructive distillation of wood to give methanol is the root of its common name—*wood alcohol*.
- Pervaporisation is a method for the separation of mixtures of liquids by partial vaporisation through a non-porous membrane.
- Dry distillation, despite its name, is not truly distillation, but rather a chemical reaction known as pyrolysis in which solid substances are heated in a strongly reducing atmosphere and any volatile fractions are collected.
- Extractive distillation is defined as distillation in the presence of a miscible, high boiling, relatively non-volatile component, the solvent, that forms no azeotrope with the other components in the mixture.
- Flash evaporation (or partial evaporation) is the partial vaporisation that occurs when a saturated liquid stream undergoes a reduction in pressure by passing through a throttling valve or other throttling device. This process is one of the simplest unit operations.
- Freeze distillation is an analogous method of purification using freezing instead of evaporation. It is not truly distillation, and does not produce products equivalent to distillation. This process is used in the production of ice-beer and ice-wine to increase ethanol and sugar content, respectively.
- Codistillation is distillation which is performed on mixtures in which the two compounds are not miscible.

Industrial Distillation

Continuous Distillation: Continuous distillation, a form of distillation, is an ongoing separation in which a mixture is continuously (without interruption) fed into the process and separated fractions are removed continuously as output streams as time passes during the operation.

A distillation is the separation or partial separation of a liquid feed mixture into components or fractions by selective boiling (or evaporation) and condensation. A distillation produces at least two output fractions. These fractions include at least one volatile distillate fraction, which has boiled and been separately captured as a vapour condensed to a liquid, and practically always a bottoms fraction, which is the least volatile residue that has not been sèparately captured as a condensed vapour.

An alternative to continuous distillation is batch distillation, where the mixture is added to the unit at the start of the distillation, distillate fractions are taken out sequentially in time (one after another) during the distillation, and the remaining bottoms fraction is removed at the end.

Because each of the distillate fractions are taken out at different times, only one distillate exit point (location) is needed for a batch distillation and the distillate can just be switched to a different receiver, a fraction-collecting container. Batch distillation is often used when smaller quantities are distilled. In a continuous distillation, each of the fraction streams is taken simultaneously throughout operation; therefore, a separate exit point is needed for each fraction. In practice when there are multiple distillate fractions, each of the distillate exit points are located at different heights on a fractionating column. The bottoms fraction can be taken from the bottom of the distillation column or unit, but is often taken from a reboiler connected to the bottom of the column.

Each fraction may contain one or more components (types of chemical compounds). When distilling crude oil or a similar feedstock, each fraction contains many components of similar volatility and other properties. Although it is possible to run a small scale or laboratory continuous distillation, most often continuous distillation is used in a large-scale industrial process.

Industrial Application

Distillation is one of the unit operations of chemical engineering. Continuous distillation is used widely in the chemical process industries where large quantities of liquids have to be distilled. Such industries are the natural gas processing, petrochemical production, coal tar processing, brewing, liquified air separation, hydrocarbon solvents production and similar industries, but it finds its widest application in petroleum refineries. In such refineries, the crude oil feedstock is a very complex multicomponent mixture that must be separated and yields of pure chemical compounds are not expected, only groups of compounds within a relatively small range of boiling points, which are called fractions. That is the origin of the name fractional distillation or fractionation. It is often not worthwhile separating the components in these fractions any further based on product requirements and economics.

Principle

The principle for continuous distillation is the same as for normal distillation: when a liquid mixture is heated so that it boils, the composition of the vapour above the liquid differs from the liquid composition. If this vapour is then separated and condensed into a liquid, it becomes richer in the lower boiling component(s) of the original mixture.

This is what happens in a continuous distillation column. A mixture is heated up, and routed into the distillation column. On entering the column, the feed starts flowing down but part of it, richer in lower boiling component(s), vaporises and rises. However, as it rises, it cools and while part of it continues up as vapour, some of it (enriched in the less volatile component) begins to descend again.

Image 2 depicts a simple continuous fractional distillation tower for separating a feed stream into two fractions, an overhead distillate product and a bottoms product. The "lightest" products (those with the lowest boiling point or highest volatility) exit from the top of the columns and the "heaviest" products (the bottoms, those with the highest boiling point) exit from the bottom of the column. The overhead stream may be cooled and condensed using a water-cooled or air-cooled condenser. The bottoms reboiler may be a steam-heated or hot oil-heated heat exchanger, or even a gas or oil-fired furnace.

In a continuous distillation, the system is kept in a steady state or approximate steady state. Steady state means that quantities related to the process do not change as time passes during operation. Such constant quantities include feed input rate, output stream rates, heating and cooling rates, reflux ratio, and temperatures, pressures, and compositions at every point (location). Unless the process is disturbed due to changes in feed, heating, ambient temperature, or condensing, steady state is normally maintained. This is also the main attraction of continuous distillation, apart from the minimum amount of (easily instrumentable) surveillance; if the feed rate and feed composition are kept constant, product rate and quality are also constant. Even when a variation in conditions occurs, modern process control methods are commonly able to gradually return the continuous process to another steady state again.

Since a continuous distillation unit is fed constantly with a feed mixture and not filled all at once like a batch distillation, a continuous distillation unit does not need a sizable distillation pot, vessel, or reservoir for a batch fill. Instead, the mixture can be fed directly into the column, where the actual separation occurs. The height of the feed point along the column can vary on the situation and is designed so as to provide optimal results.

A continuous distillation is often a fractional distillation and can be a vacuum distillation or a steam distillation.

Design and Operation

Design and operation of a distillation column depends on the feed and desired products. Given a simple, binary component feed, analytical methods such as the McCabe-Thiele method or the Fenske equation can be used to assist in the design. For a multi-component feed, computerised simulation models are used both for design and subsequently in operation of the column as well.

Modelling is also used to optimise already erected columns for the distillation of mixtures other than those the distillation equipment was originally designed for.

When a continuous distillation column is in operation, it has to be closely monitored for changes in feed composition, operating temperature and product composition. Much of these tasks are performed these days using advanced computer control equipment.

Column Feed

The column can be fed in different ways. If the feed is from a source at a pressure higher than the distillation column pressure, it is simply piped into the column. Otherwise, the feed is pumped or compressed into the column. The feed may be a superheated vapour, a saturated vapour, a partially vaporised liquid-vapour mixture, a saturated liquid (i.e., liquid at its boiling point at the column's pressure), or a sub-cooled liquid. If the feed is a liquid at a much higher pressure than the column pressure and flows through a pressure let-down valve just ahead of the column, it will immediately expand and undergo a partial flash vaporisation resulting in a liquid-vapour mixture as it enters the distillation column.

Improving Separation

Although small size units, mostly made of glass, can be used in laboratories, industrial units are large, vertical, steel cylinders known as "distillation towers" or "distillation columns". To improve the separation, the tower is normally provided inside with horizontal plates or trays, or the column is packed with a packing material. To provide the heat required for the vaporisation involved in distillation and also to compensate for heat loss, heat is most often added to the bottom of the column by a *reboiler*, and the purity of the *top product* can be improved by recycling some of the externally condensed top product liquid as reflux.

Depending on their purpose, distillation columns may have liquid outlets at intervals up the length of the column as shown below:

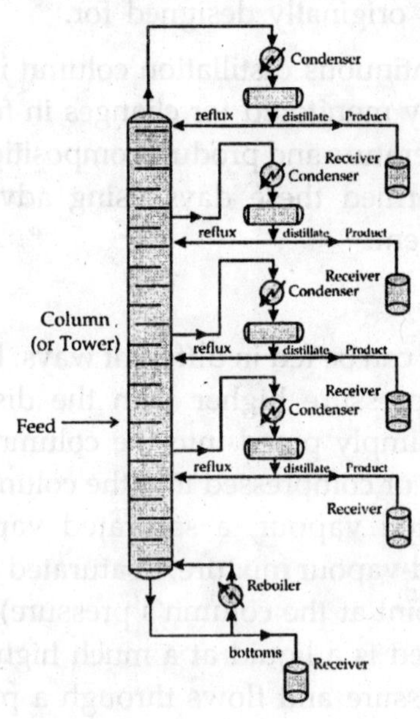

figure

Reflux

Large-scale industrial fractionation towers use reflux to achieve more efficient separation of products. Reflux refers to the portion of the condensed overhead liquid product from a distillation tower that is returned to the upper part of the tower. Inside the tower, the downflowing reflux liquid provides cooling and partial condensation of the upflowing vapours, thereby increasing the efficacy of the distillation tower. The more reflux that is provided, the better is the tower's separation of the lower boiling from the higher boiling components of the feed. A balance of heating with a reboiler at the bottom of a column and cooling by condensed reflux at the top of the column maintains a temperature gradient (or gradual temperature difference) along the height of the column to provide good conditions for fractionating the feed mixture.

Changing the reflux (in combination with changes in feed and product withdrawal) can also be used to improve the separation properties of a continuous distillation column while in operation (in contrast to adding plates or trays, or changing the packing, which would, at a minimum, require quite significant downtime).

Plates or trays

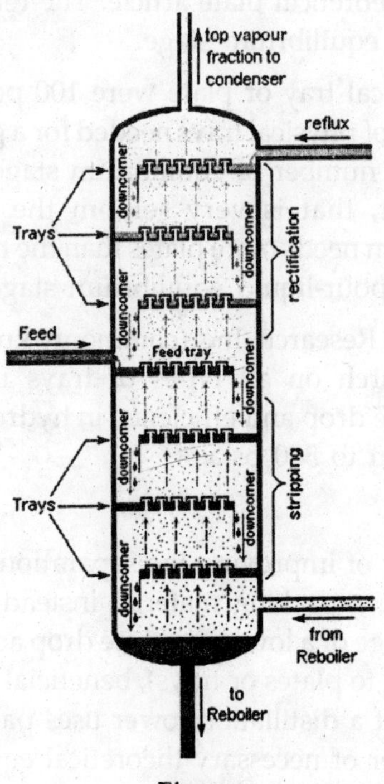

Figure

Distillation towers use various vapour and liquid contacting methods to provide the required number of equilibrium stages. Such devices are commonly known as "plates" or "trays". Each of these plates or trays is at a different temperature and pressure. The stage at the tower bottom has the highest pressure and temperature.

Progressing upwards in the tower, the pressure and temperature decreases for each succeeding stage. The vapour-liquid equilibrium for each feed component in the tower reacts in its unique way to the different pressure and temperature conditions at each of the stages. That means that each component establishes a different concentration in the vapour and liquid phases at each of the stages, and this results in the separation of the components. Some example trays are depicted in image 4. A more detailed, expanded image of two trays can be seen in the theoretical plate article. The reboiler often acts as an additional equilibrium stage.

If each physical tray or plate were 100 per cent efficient, than the number of physical trays needed for a given separation would equal the number of equilibrium stages or theoretical plates. However, that is very seldom the case. Hence, a distillation column needs more plates than the required number of theoretical vapour-liquid equilibrium stages.

Fractionation Research, Inc. (commonly known as FRI) has performed research on all types of trays measuring their capacity, pressure drop and efficiency in hydrocarbon systems from full vacuum to 500 psia.

Packing

Another way of improving the separation in a distillation column is to use a packing material instead of trays. These offer the advantage of a lower pressure drop across the column (when compared to plates or trays), beneficial when operating under vacuum. If a distillation tower uses packing instead of trays, the number of necessary theoretical equilibrium stages is first determined and then the packing height equivalent to a theoretical equilibrium stage, known as the height equivalent to a theoretical plate (HETP), is also determined. The total packing height required is the number theoretical stages multiplied by the HETP.

This packing material can either be random dumped packing such as Raschig rings or structured sheet metal. Liquids tend to wet the surface of the packing and the vapours pass across this wetted surface, where mass transfer takes place. Unlike conventional tray distillation in which every tray represents a separate point of vapour-liquid equilibrium, the vapour-liquid equilibrium curve in a packed column is continuous.

However, when modelling packed columns it is useful to compute a number of theoretical plates to denote the separation efficiency of the packed column with respect to more traditional trays. Differently shaped packings have different surface areas and void space between packings. Both of these factors affect packing performance.

Another factor in addition to the packing shape and surface area that affects the performance of random or structured packing is liquid and vapour distribution entering the packed bed. The number of theoretical stages required to make a given separation is calculated using a specific vapour to liquid ratio. If the liquid and vapour are not evenly distributed across the superficial tower area as it enters the packed bed, the liquid to vapour ratio will not be correct in the packed bed and the required separation will not be achieved.

The packing will appear to not be working properly. The height equivalent to a theoretical plate (HETP) will be greater than expected. The problem is not the packing itself but the mal-distribution of the fluids entering the packed bed. Liquid mal-distribution is more frequently the problem than vapour. The design of the liquid distributors used to introduce the feed and reflux to a packed bed is critical to making the packing perform at maximum efficiency. Methods of evaluating the effectiveness of a liquid distributor can be found in references. Considerable work as been done on this topic by Fractionation Research, Inc.

Overhead System Arrangements

In many cases, the tower overhead is not easily condensed totally and the reflux drum must include a vent gas outlet stream. In yet other cases, the overhead stream may also contain water vapour because either the feed stream contains some water or some steam is injected into the distillation tower (which is the case in the crude oil distillation towers in oil refineries). In those cases, if the distillate product is insoluble in water, the reflux drum may contain a condensed liquid distillate phase, a condensed water phase and a non-condensible gas phase, which makes it necessary that the reflux drum also have a water outlet stream.

Examples

Continuous Distillation of Crude Oil: Petroleum crude oils contain hundreds or more different hydrocarbon compounds: paraffins, naphthenes and aromatics as well as organic sulphur compounds, organic nitrogen compounds and some oxygen containing hydrocarbons such as phenols. Although crude oils generally do not contain olefins, they are formed in many of the processes used in a petroleum refinery.

The crude oil fractionator does not produce products having a single boiling point, rather, it produces fractions having boiling ranges. For example, the crude oil fractionator produces an overhead fraction called "naphtha" which will become a gasoline component after it is further processed through a catalytic hydrodesulphuriser to remove sulphur and a catalytic reformer to reform its hydrocarbon molecules into more complex molecules with a higher octane rating value.

The naphtha "cut", as that fraction is called, has very many different hydrocarbon compounds. Therefore it has an "initial" boiling point of about 35°C and a "final" boiling point of about 200°C that is what is meant by the "boiling range" of each "cut" produced in the fractionating columns. At some distance below the overhead, the next "cut" is withdrawn from the side of the column and it is usually the jet fuel cut also known as a kerosene cut. It also contains very many different hydrocarbons and the boiling range of that cut is from an initial boiling point of about 150°C to a final boiling point of about 270°C. The next cut further down the tower is the diesel oil cut with a boiling range from about 180°F to about 315°C. Note the overlap of boiling range between any cut and the next cut because the distillation separations are not perfectly sharp.

After these come the heavy fusel oil cuts and finally the bottoms product, with very wide boiling ranges. All these cuts are processed further in subsequent refining processes.

Distillation in Food Processing

Distilled Beverage: A distilled beverage is a liquid preparation meant for consumption containing ethyl alcohol (ethanol) purified by distillation from a fermented substance such as fruit, vegetables, or grain. The word *spirits* generally refers to distilled beverages low in sugars and containing at least 35 per cent alcohol by volume. Ginger Wine, baijiu, gin, vodka, rum, whisky, brandy, absinthe, tequila, and traditional German schnapps are types of spirits. Distilled beverages with added flavourings and a relatively high sugar content such as Grand Marnier, Frangelico and American style schnapps are generally referred to as liqueurs. The term *liquor* may mean spirits; spirits and liqueurs; or all alcoholic beverages, including wine, sake, beer, and mead.

Distillation History

Beer and wine were historically limited to a maximum alcohol content of about 15 per cent by volume, beyond which yeast is adversely affected and cannot ferment. Alcohol levels higher than 15 per cent have historically been obtained in a number of ways.

Wine heated in an animal bladder draws out water and

leaves alcohol behind (the bladder has a natural property which removes water), but there is no evidence this method was used before modern times.

The first evidence of true distillation comes from Babylonia and dates from the fourth millennium BC. Specially shaped clay pots were used to extract small amounts of distilled alcohol through natural cooling for use in perfumes, however it is unlikely this device ever played a meaningful role in the history of the development of the still. By the 3rd century AD there is evidence that alchemists in Alexandria, Egypt, used distillation to produce alcohol for sublimation and for colouring metal.

Central Asia

Freeze distillation, the "Mongolian still", are known to have been in use in Central Asia as early as the 7th century AD. The first method involves freezing the alcoholic beverage and removing water crystals. The freezing method had limitations in geography and implementation and thus did not have widespread use, but remained in limited use, for example during the American colonial period applejack was made from cider using this method.

Middle East

The development of the still with cooled collector—necessary for the efficient distillation of spirits without freezing—was an invention of Persian alchemists in the 8th or 9th centuries. In particular, Geber (Jabir Ibn Hayyan, 721-815) invented the alembic still; he observed that heated wine from this still released a flammable vapour, which he described as "of little use, but of great importance to science". Not much later Al-Razi (864-930) described the distillation of alcohol and its use in medicine. By that time, distilled spirits had become fairly popular beverages: the poet Abu Nuwas (d. 813) describes a wine that "has the colour of rainwater but is as hot inside the ribs as a burning firebrand". The terms "alembic" and "alcohol", and possibly the metaphors "spirit" and *aqua vitae* ("lifewater") for the distilled product, can be traced to Middle Eastern alchemy.

Names like "life water" have continued to be the inspiration for the names of several types of beverages, like Gaelic whisky, French eaux-de-vie and possibly vodka. Also, the Scandinavian akvavit spirit gets its name from the Latin phrase *aqua vitae*.

Medieval Europe

Distilled alcohol beverages first appeared in Europe in the mid-12th century among alchemists, who were more interested in medical "elixirs" than making gold from lead. It first appears under the name *aqua ardens* (burning water) in the *Compendium Salerni* from the medical school at Salerno. The recipe was written in code, suggesting it was kept a secret. Taddeo Alderotti in his *Consilia medicinalis* referred to the "serpente" which is believed to have been the coiled tube of a still.

Paracelsus gave alcohol its modern name, taking it from the Arabic word which means "finely divided", in reference to what is done to wine. His test was to burn a spoonful without leaving any residue. Other ways of testing were to burn a cloth soaked in it without actually harming the cloth. In both cases, to achieve this effect the alcohol had to have been at least 95 per cent, close to the maximum concentration attainable through fractional distillation.

Claims on the origins of specific beverages are controversial, often invoking national pride, but they are plausible after the 12th century when Irish whiskey, German Hausbrand and German brandy can all be safely said to have arrived. These beverages would have had much lower alcohol content than the alchemists' pure distillations (around 40 per cent by volume), and were likely first thought of as medicinal elixirs. Consumption of distilled beverages rose dramatically in Europe in and after the mid 14th century, when distilled liquors were commonly used as remedies for the Black Death. Around 1400 it was discovered how to distil spirits from wheat, barley, and rye beers; even sawdust was used to make alcohol, a much cheaper option than grapes. Thus began the "national" drinks of Europe: jenever (Belgium and the Netherlands), gin (England), schnapps (Germany), akvavit (Scandinavia), vodka (Russia and Poland), rakia (the Balkans). The actual names only emerged in the 16th century but the drinks were well known prior to that date.

Modern Distillation

The actual process of distillation itself has not changed since the 8th century. There have, however, been many changes in both the methods by which organic material is prepared for the still and in the ways the distilled beverage is finished and marketed.

Knowledge of the principles of sanitation and access to standardised yeast strains have improved the quality of the base ingredient; larger, more efficient stills produce more product per square foot and reduce waste; ingredients such as corn, rice, and potatoes have been called into service as inexpensive replacements for traditional grains and fruit. Chemists have discovered the scientific principles behind ageing, and have devised ways in which ageing can be accelerated without introducing harsh flavours.

Modern filters have allowed distillers to remove unwanted residue and produce smoother finished products. Most of all, marketing has developed a worldwide market for distilled beverages among populations which in earlier times did not drink spirits.

Microdistilling is a trend that began to develop in the United States following the emergence and immense popularity of microbrewing and craft beer in the last decades of the 20th century. It is specifically differentiated from megadistilleries in the quantity, and arguably quality, of output.

In most jurisdictions, including those which allow unlicensed individuals to make their own beer and wine, it is illegal to distil beverage alcohol without a license.

Chemical Profile

A distilled beverage is typically manufactured by distillation, ageing if applicable and dilution to the set percentage of alcohol.

Distillation is done at least twice, due to the chemistry involved. Copper is typically used as a chemically near-inert metal for the equipment. However, it is still very much a transition metal catalyst, and catalyses the formation of poisonous and harmful by-products, such as urethane. Removal of these is necessary and warrants a second distillation step. Most "coloured" alcohols are distilled in a batch process, but continuous processes are found in the production of flavourless vodka and similar drinks.

After distillation, the alcohol may be aged in traditional oak casks. Whiskey, for example, is aged at 77 per cent. Dilution is done to attain the standard percentage, from 30 to 80 per cent. The (arbitrary) percentage of 40 per cent is the most common "standard". However, a lower percentages such as 38 per cent may make the drink more palatable. Also people often mix water into the drink to suit their tastes.

The final drink contains water, alcohol, fusel oils, and flavouring compounds. In some cases, sugar is added. Fusel alcohols are higher alcohols than ethanol, are mildly toxic, and have a strong, disagreeable smell and taste. Fusels in moderate quantities are considered to be essential parts of the taste profile of flavoured drinks such as whiskey and cognac. In drinks intended to be relatively flavourless (such as vodka), they are defects. Incompetently distilled drinks also contain distillation heads, which are poisonous in large amounts and consist mostly of methanol and foul-smelling by-products of fermentation.

3

Chemical Industry

There are five principal methods whereby pressure vessels may be constructed, viz. by casting, soldering and brazing, riveting, welding and hollow forging. Oust vessels are usually of comparatively small dimensions, and for moderate pressures. Soldering and brazing methods are confined to small non-ferrous vessels. Riveted construction follows the same general principles which have been developed for steam-boiler practice, but due to recent marked advances in the art of fusion welding, riveting is fast being replaced by welded construction carried out under careful control.

Hollow forging is employed for vessels for the very highest pressures and severe operating conditions, as for example in the hydrogenation of coal, and the synthesis of ammonia and methyl alcohol.

In Great Britain the only official regulation for the actual design of steam boilers and pressure vessels is contained in the appropriate sections of the Board of Trade Instructions as to the survey of Passenger Steamships. These instructions are in respect of the general principles of design, and rules for computing the dimensions of the various stressed parts of pressure vessels, such as boilers, steam-pipes, air receivers and Diesel engine starting-bottles.

There are no corresponding rules for vessels for land service, but in so far as steam boilers are concerned, the Board of Trade rules are usually applied because all boiler explosions in this country are investigated by the Board of Trade under the Boiler Explosions Acts of 1882 and 1890.

These Acts contain clauses to the effect that *(a)* notice of all boiler explosions with full particulars must be sent to the Board of Trade within 24 hours, and *(b)* the word "Boiler" is defined as any closed vessel used for generating steam, for heating water or any other liquid, or into which steam is admitted for heating, steaming, boiling or other similar purposes.

The Factories Act of 1937 requires that every steam boiler for the generation of steam must be equipped with certain fittings, that boiler and accessories must be properly maintained, and that such boilers must be examined by a competent person every fourteen months. In practice, such examinations are carried out through one or other of the Boiler Insurance Companies.

The Factories Act of 1937 extends the principle of steam-boiler inspection (as required by the 1901 Factory and Workshops Act) to include practically every other type of steam-pressure vessel and air receiver, in that such vessels must be examined by a competent person once every twenty-six months.

There are certain other statutory requirements in regard to pressure vessels in works designated as Chemical Works under the Chemical Works Regulations, 1922 (S.R.O.No. 731, of 1922). Para. 5 requires that every still and every closed vessel in which gas is evolved or into which gas is passed and in which the pressure is liable to rise to a dangerous degree, shall have attached to it and maintained in proper condition a proper safety valve or other equally efficient means to relieve the pressure. Para. 6 of these Regulations deals with the provision of breathing apparatus and lifebelts, and this leads naturally to paras. 7, 8, 9, 19, 20, which govern the entry of workpeople into vessels, chambers, structures of any sort, and any room or place in which there is reason to apprehend the presence of a dangerous gas or the existence of an irrespirable atmosphere. These regulations are of the utmost importance in regard to the design of plant and methods used in construction in regard to access, cleaning and repairs.

From the aspect of working pressure it is convenient to divide pressure vessels for the chemical industry into two groups.

1. Vessels for service at pressures not higher than about 1000 lbs. per sq. in. and wherein the "thin-cylinder" formula may be applied. This pressure limit may be raised if high-tensile alloy steels are used, since in such cases this working stress may be considerably higher than plain carbon steels and consequently the wall thickness is sufficiently thin for the " thin-cylinder " theory to be applied.
2. Vessels for service at pressures higher than 1000 lbs. per sq. in. and wherein the "thick-cylinder" or Lame's theory may be applied.

Each of these two groups may be again subdivided on the basis of the working temperature of the pressure vessel concerned, since if the working temperature of the stressed parts is within the region wherein " creep " occurs, the above principles of design based upon the thin-cylinder and thick-cylinder theories must be modified accordingly. The modification consists in substituting the safe working stress at the working temperature.

A pressure vessel consists essentially of a middle and two ends. The middle portion is practically always cylindrical in shape, and the ends may be either flat, dished, or hemispherical. The principles underlying the design of the shell or cylindrical portion will first be discussed.

Vessel Shells

If a thin circular cylinder is subjected to internal fluid pressure, a tensile stress arises in the walls of the material, and this is usually described as the "hoop stress". When the thickness of the cylinder walls is small compared with the internal diameter of the cylinder, it is assumed that the hoop stress is uniformly distributed throughout the cross-section of the material.

Construction of Pressure Vessels

The general practice of riveting in chemical pressure-vessel work follows closely the technique long developed for steam-boiler construction. It is of importance to note that apart from failure by leakage a riveted joint can fail only in certain well-defined ways:

(1) Failure of rivets in shear.

(2) Tearing of plate through a line of rivet-holes.

(3) Combined tearing of plate and shearing of rivets.

(4) Lap rupt 111 × 3.

These four possibilities form the basis for the design of riveted joints:

(1) Is accounted for by providing an adequate section of rivet-material in shear;

(2) By providing an adequate section of plate to resist the tearing process;

(3) By taking account of the weakest section of the plate and adding to its strength, the resistance of the rivets which would have to shear in order to allow the tearing to take place; and

(4) By providing adequate material between rows of rivet holes and between the holes and the plate edges.

In general, the efficiency of a riveted joint may be defined as the ratio between the minimum calculated resistance to failure, and the ultimate tensile strength of the solid plate, and this ratio is normally used in the calculation of the working pressure. It should be noted that it is the ultimate tensile strength of tin; plate material which provides the criterion of efficiency are riveted joints, and, so far as pressure vessels are concerned, this criterion has been found by experience to be entirely satisfactory.

Riveted construction, however, does not permit of good design in the case of many chemical pressure; vessels: first, because of the interference of riveted joints with a smooth interior; second, the stresses set up in the shell and rivets in the riveting process often lead to rapid local corrosion; third, because many metals do not lend themselves to riveted work.

The following rules on riveting of boilers are taken from the Board of Trade Rules by permission of the Controller of I.T.M. Stationery Office, and arc given as a guide; to good practice in the riveting of pressure vessels.

Note that these rules apply to 'wlired " pressure vessels, and for vessels of the unfired type the working pressure may be increased by 10 per cent above that found by these formulae for fired vessels.

Distances between Rows of Rivets and between Rivets and Plate Edges:

(a) The clear space between a rivet hole and the edge of a plate should not be less than the diameter of the rivet hole, i.e. the centre of the rivet hole should be at least U diameters distant from the edge of the plate.

(b) In zigzag riveted joints, whether lapped or fitted with butt straps, in which there is an equal number of rivets in each row.

(c) In chain-riveted joints, whether lapped or fitted with butt straps, in which there is an equal number of rivets in each row, the distance between the rows should not be less than 2*d*.

(d) In zigzag riveted joints in which the number of rivets in a row is one-half of the number in an adjacent row.

(e) In chain-riveted joints in which the number of rivets in a row is one-half of the number in an adjacent row. The distance between rows in which there are the full number of rivets should be not less than $2d$. Where d is diameter of the rivet holes.

Methods of Calculating the Strength of Riveted Joints: The percentage of strength of a riveted joint is found from the following formulae (I), (II), (III): (I) and (II) are applicable to any type of joint; (III) is applicable only to joints in which the number of rivets in inner rows is double that in the outer row. The lowest value given by the application of these formulae is to be taken as the percentage of strength of the joint.

(I) Percentage of strength of plate at joint as compared with solid plate – $100\ (p - d)\ P$

(II) Percentage of strength of rivets as compared with the solid plate – $100\ (S_2 \times a \times n \times C) \sim S_1\ x{\sim}p \approx x{\sim}T$

(III) Percentage of combined strength of the plate at the inner row of rivet holes and of the rivets in the outer row = $100(p–2d), 100(S_2 \times a \times C) \sim p\ S_1 \times p \times T$

where p = pitch of rivets at outer rows in in.,

d = diameter of rivet holes in in;,

a ~ sectional area of one rivet in sq. in.,

n = number of rivets which arc fitted in the pitch p,

T = Thickness of plate in in.,

0 = 1.0 for rivets in single shear as in lap joints,

C ~ 1.875 for rivets in double shear as in double butt-strapped joints,

S_1 = minimum tensile strength of plates in tons/ sq. in.,

S_2 = shearing strength of rivets, which is taken generally to be 23 tons/sq. in., and may be 85 per cent, of the minimum tensile strength of the rivet bars.

Circumferential Seams

(a) The riveting of the seams joining the end plates to the cylindrical shell shall be not less than 42 per cent, of that of the solid plate. Where the shell plates exceed 1 in. in thickness the seams connecting the shell plates to the end plates are to be double riveted.

(b) The circumferential seam at or near the middle of the length of single-ended boilers should have a strength of joint not less than 60 per cent, of the solid plate. The inner circumferential seams of double-ended boilers should have a strength of joint not less than 62 per cent, of the solid plate. In any case, there shall be three rows of rivets.

(c) The circumferential Seams in the shell of a vertical boiler shall be not less in strength than 42 per cent, that of the solid plate. When the seams are not complete circles, they shall be double riveted.

Welded Construction of Pressure Vessels

Before discussing the various rules for welded construction of pressure vessels it is considered desirable to describe in detail the various methods of welding followed by notes on design.

Fabrication Pressure Vessels

The welding processes used in the manufacture of pressure vessels, fabricated plate work and accessories may be broadly classified as follows:

(1) Forge and hammer welding.
(2) Electric resistance butt welding.
(3) Atomic hydrogen process.
(4) Metallic arc fusion welding.
(5) Acetylene welding.

Forge and Hammer Welding

The forge and hammer welding method of pressure-vessel construction is closely allied to the oldest known method of welding, and has been used for generations in the manufacture of furnace and flue tubes for Lancashire and Cornish boilers. This method has recently been developed, chiefly in Germany, for the fabrication of large pressure vessels, and the following description of the technique adopted by Messrs. Thyssen of Mulheim Ruhr is of interest in this connection.

The pressure vessels are made from standard qualities of steel plate, and for the cylindrical portion, the plate is first put through rolls to obtain circular shape. The welding of the longitudinal seam is carried out with a water gas flame of a reducing nature, and the edges of the plate being welded are worked by means of rolling at welding temperature. After completion of the longitudinal seam, this is then closely examined, and if satisfactory the vessel is annealed and an internal hydraulic; pressure of 50 per cent, above the desired working pressure is applied. The ends of the vessel are then closed in, and manhole and other openings fitted.

Upon completion, the vessel is then highly stressed by application of an hydraulic pressure of three or four times the working pressure, and finally the vessel is heat-treated to relieve internal stresses.

During the past twelve years, Messrs. Thyssen have supplied more than 1250 such vessels, chiefly in the form of steam drums for water-tube boilers, and in the majority of eases the working pressures were in excess of 500 lbs. per sq. in.

Electric Resistance Butt Welding

For joining of parts of tubular structures this method offers many advantages, and it can be used with great facility in the construction of elements of tubular heaters and heat exchangers. It has the advantage that when the weld is complete, there has been no addition of metal from any outside source. The method of operation is to place the tubes to be welded one into each of two clamps which form the electrodes of the machine. One of the clamps is stationary and forms part of the rigid structure of the machine, and the other forms a sliding head attached to a mechanically operated ram capable of either slow or rapid movement.

Current flows between the two clamps through the tubes, and when the ends of the tubes meet a resistance is set up due to discontinuity of the structure. This resistance raises the temperature of the metal at the ends of the tube, a certain amount of metal is flashed off, and when at the right plasticity an upset or forge blow is imparted to form the weld.

The current used is supplied through a transformer, the secondary windings being tapped to give voltages ranging between 5-8 volts. The cost of equipment and of power consumption in this process limits the size of work to tubular structures in which the cross-sectional area to be welded does not exceed about 30 sq. ins.

Atomic Hydrogen Process

The use of atomic hydrogen is one of the later developments in the field of welding and appears to have certain advantages. Although an electric arc is used, its application is indirect, and up to the present covered electrodes have not been utilised to any extent. It may be looked upon as supplementary to metallic arc welding rather than as a competitive process.

An arc is maintained by an alternating current between two tungsten wire electrodes and a stream of hydrogen gas is introduced around the arc. The temperature of the electric arc is sufficiently high to cause dissociation of the hydrogen molecules, and the atomic hydrogen on leaving the zone of the arc recombines to form again hydrogen in the molecular state. The latter process is a reversal of the heat input from the electric arc, energy is given up, and provides the heat required for welding.

Such a process may appear to be a complicated method of providing the heat required for welding, but by this means it is possible to produce a flame of high temperature composed of a single gas without the usual combination with oxygen, whilst, in addition, the welding can be carried out in an envelope of burning hydrogen so that the metal constituting the weld can be maintained free from atmospheric contamination without a slag coating.

Metallic Arc Fusion Welding

In view of the importance of this method in the fabrication of pressure vessels, the subject will be dealt with at length.

When the early development of welded pressure vessels began, electric arc welding offered great possibilities. An automatic machine had been devised for welding cylindrical containers made from thin steel plates. This machine worked well with bare wire electrodes, but the class of weld which could be made was unsuitable for pressure vessels since metal deposited in the weld from bare electrodes contains a number of major defects, such as incomplete fusion with parent metal, slag inclusions and porosity.

Apart from the presence of such defects, the use of the bare electrode results in considerable atmospheric contamination of the weld metal, this contamination being exaggerated by the instability of the arc when using electrodes of this type. The atmospheric contamination causes oxide and nitride inclusions in the weld metal which tend to make the metal brittle, even if present only in minute quantities.

The weld metal must have physical and chemical properties similar to those of the parent metal, and means must be provided to replace in the weld metal the carbon, silicon and manganese lost due to volatilisation of these elements in the arc. Coated electrodes, therefore, have to be used, and machines have to be developed capable of feeding such electrodes without stopping and starting at short intervals for electrode replacement. The manufacturer has to consider vessels up to 80 ft. in length, in which long longitudinal seams are a necessity. Means have been found for automatically feeding long-coated electrodes to provide a continuous weld. The advantage of such a method of working, combined with constant, mechanically regulated short arc conditions, will be easily appreciated. Weld metal can be laid down within definite temperature limits so that the class of structure can be controlled throughout the welding operation.

The electrode consists of two portions, the steel wire and the coating attached thereto.

The wire is delivered in coils and the first operation involves the cleaning of its surface and straightening and cutting it into the 8-ft. lengths suitable for the automatic welding machines. In the second operation projections are stamped by the machine on to the wire at a regular pitch.

These projections just protrude through the electrode mating, so that electrical contact can be made with the contactors of the feeding device on the automatic machines. In order that the automatic, welding process is not interrupted, the electrodes are made so that the one can lit into the other, thus providing what is essentially a continuous electrode. After the stamping operation, the wire is fed to the machine which extrudes the coating, and applies a carefully controlled thickness of the coating to the wire.

The Welding Technique: The deposition of weld metal by means of the metallic arc involves a close study of three main variables:

(a) The electrical input to the arc;

(b) The speed at which the electrode travels along the seam being welded; and

(c) The shape of the groove formed by the abutting plates.

The quality of weld metal deposited from an electrode depends largely on the amperage and voltage of the arc. The second variable may be considered as follows: with given electrical conditions and an excessive speed of travel of the electrode along the seam, the heat generated in the vicinity of the arc will not be sufficient to fuse the sides of the groove being welded, and at the junction between the weld metal and the plate there will remain a thin film of non-metallic material, for instance slag, which will constitute a definite weakness at the joint. Time is naturally an important factor in raising the temperature of the parent metal to the degree necessary.

If the speed of the electrode is too low, three conditions can follow. First, the run of weld metal deposited is of too great a thickness so that subsequent recrystallisation by the succeeding run is confined to a relatively small depth of the metal. Secondly, the heat generated in the vicinity of the arc causes excessive fusion of the plate, and pockets enclosing slag are formed. Thirdly, the slag arising from the electrode coating, instead of being controlled to freeze behind the arc, will flow in front of the electrode and become overlapped by the weld metal.

Having once, therefore, determined the electrical input to the arc for good-quality weld metal, the speed of travel of the electrode must be considered in relation to the shape of the groove in the plate.

A groove, having the U-shaped section is almost universal for heavy plate because the change in the width of the groove after each layer of metal has been deposited is almost negligible, and because the almost vertical line of fusion is admirably suited to X-ray inspection.

The minimum dimensions of the groove arc controlled by the electrode used. The narrower the groove, the less the amount of weld metal necessary and the cheaper the fabrication. The factor of primary importance which determines the minimum width of groove is the first layer of weld metal deposited. If the groove is too pointed at the root of the weld, inadequate penetration at the base of the groove will follow.

Regarding size of electrodes, it might appear that the most economical procedure would be to use electrodes of large cross-section, thereby filling a deep groove in two or three runs. It sometimes happens in such cases that cheapness and speed do not go hand in hand with the soundness of weld in the manufacture of pressure vessels.

The four major considerations are:

(a) Distortion of the vessel due to unbalanced contraction where large amounts of weld-metal are laid down;

(b) The retention of a coarse columnar structure, because of the restricted depth to which the refining effect of the superposed run of weld-metal can penetrate;

(c) The trapping of slag in the weld-metal due to freezing before the slag has floated to the surface of the weld run;

(d) High current density necessary with large electrodes produces excessive porosity.

The multilayer weld is most attractive. The comparatively thin layers of weld-metal permit the full flotation of slag before freezing, and with suitable electrode coatings a protective slag with a low specific gravity can be ensured. Whilst the current density can be maintained sufficiently high to ensure good fluidity of weld-metal and complete fusion, the smaller section electrodes do not necessitate current densities of the class which bring about porosity.

Attention must also be paid to the disturbed area in the parent metal immediately on each side of the weld. There are many practical difficulties which prevent the normalising of the finished drum. It is not possible to heat the finished drum to a temperature of the order of 900°C to recrystallise the material.

In addition, considering the weld-metal, this recrystallisation would be undesirable because the rate at which the whole mass of the drum could be cooled through the upper 400°C. of the temperature curve would be such that a grain structure of very much coarser type would result in the heat-treated weld itself, and the laminar structure of the plate would lose the fibrous nature which makes it particularly suited to carry circumferential stresses, such as the hoop stress, etc. to which the drum is subjected in service. The only theoretical reason upon which recrystallisation in such a vessel can be advocated is to remove the disturbed area in the parent metal immediately adjacent to the point at which fusion between the weld-metal and parent metal takes place.

In discussing the structure of the parent metal adjacent to the weld, reference has been made to the important question of heat treatment. In considering this question, it is desirable to bear in mind the class of vessel which may have to be handled. Any attempt to normalise the large shells, now fabricated at a temperature above the critical range of about 900°C, would result in serious deformation because of the plasticity of the metal at that temperature. For metallurgical reasons, rapid cooling through the temperature range between 900° and 600°C would be necessary, but it would be impossible with hollow cylindrical vessels because of the danger of unequal stresses due to contraction and the considerable heat contained in the large mass of metal presented by the vessel.

The heat treatment to which vessels are subjected consists of bringing the metal to a temperature of 600°C, and maintaining it at that temperature for a period of 1 hours for each inch of thickness of the weld, the vessel being allowed to cool slowly afterwards in suitable screens. There can be no doubt that low-temperature stress relief treatment is beneficial. Suggestions have been made from time to time that a higher stress relief temperature ranging between 650° and 700°C should be employed. Whilst there is no doubt that the plasticity of the metal will increase with temperature, it has to be remembered that within this temperature range the carbide forms spheroidal aggregates at a relatively rapid rate, and any small benefit would only be gained at the expense of the structure of the metal.

Non-destructive Tests: In welded pressure vessels, test plates can easily be provided at each end of the longitudinal seam, these plates being tacked on prior to welding and butt-welded as part of the longitudinal seam of the vessel.

A welded cylinder with the test plates in position. After

the welding is completed and the plates removed, they are given a controlled and recorded heat treatment exactly similar to that to which the drum is subjected, and are afterwards cut up into suitable specimens for transverse tensile, bend, all-weld-metal tensile, weld-metal density, and impact tests, whilst the material is also examined microscopically.

Such plates, however, cannot be produced when welding circumferential seams, as machine-welded circumferential scams are continuous. Some means, therefore, have to be found whereby it is possible to establish that the quality of the weld-metal shown by the mechanical tests of the test plates is identical with that in the whole length of the longitudinal and circumferential seams. Examination by X-rays has proved to be the most satisfactory means for attaining this end. By radiographic inspection of every inch of the weld and of the test plates, it is possible to establish a direct comparison between the weld-metal in the test plates and the weld-metal in the vessel itself.

The radiographic inspection is considerably simpler than might be supposed. The X-rays emitted by the tube penetrate the weld from the outside of the vessel, so that the radiograph is produced on a highly sensitised film carried in a cassette secured to the inner surface of the drum along the weld. The radiograph produced in a shadow picture, and the reading of these radiographs, whilst initially difficult, soon becomes with practice comparatively easy.

The most satisfactory way of obtaining a first-class knowledge of the meaning of radiographs is that of making welds under conditions which are deliberately varied, examining them by X-rays, making sections, and afterwards comparing these with the pictures produced and studying the history of the weld.

The code of the American Society of Mechanical Engineers provides a set of radiographs which are chosen to indicate acceptable or unacceptable welds, and this comparative basis for examination has proved satisfactory since its inclusion in the Code in 1931.

Acetylene Welding

Compared with even a few years ago, the standard of oxy-acetylene welding, both in cost and quality, shows remarkable progress. The methods of making oxy-acetylene welds can be classified under the following headings:

(1) Leftward welding.

(2) Rightward welding.

(3) Upward vertical welding (double reinforcement)

Method A: Welding from one sides only.

Method B: Welding simultaneously from both sides.

Upward Vertical Welding, Method "A." (Oxy-Acetylene Process.)

Direction of Welding: The welding rod, the flame and the seam are in the same vertical plane. The sheets to be joined are arranged in a vertical position, the weld being made in a vertical position.

Direction and Position of the Flame: The tip of the blowpipe, and consequently the flame, is inclined at an angle of 30° to the horizontal.

Direction and Position of the Welding Rod: The welding rod makes an angle of 20° with the horizontal, in a direction opposite to that of the tip of the blowpipe. The end of the rod is held in the molten metal and just within the edges to be joined.

Size of the Welding Rod: The diameter should be based upon the thickness of the sheet or plate and should be equal to one-half the thickness of the material to be welded.

Movement of the Flame: After forming a small hole by melting the two edges—the hole being maintained during the welding of the seam— only a steady upward movement in the direction of the seam is necessary.

Movement of the Welding Rod: The welding rod is drawn backward and forward in the molten metal.

As compared with the methods of rightward and leftward welding, it ensures complete control of the double reinforcement, as in rightward welding, but does not give the same reducing flame protection on the reverse side of the seam. It is, therefore, not so well adapted for metals which oxidise readily at a red heat. The power of the blowpipe being less, the speed of welding is consequently less, and the cost of welding slightly higher. The advantages include the decreased importance of the gap between the edges and the small demands made upon the concentration of the welder. Penetration is much more certain than with leftward welding.

Upward Vertical Welding, Method " B"

This method requires the services of two welders, welding simultaneously on either side of the seam.

Direction of Welding: The welding rod, the flame and the seam are in the same vertical plane. The plates to be joined are arranged in a vertical position, the welds being made in an upward direction.

Direction and Position of the Flames: The tip of each blowpipe, and consequently the flame, is inclined at an angle of 30° to the horizontal.

Direction and Position of the Welding Rods: Each welding rod makes an angle of 20° with the horizontal, and in a direction opposite to that of the tip of the blowpipe. The end of the welding rod is held in the molten metal, within the edges to be joined.

Size of the Welding Rods: The diameter of the welding rods should be 10 gauge.

Movement of the Flames: As soon as the first common pool of metal has been obtained, by each welder penetrating one-half of the plate at the same level, only a steady upward movement is necessary.

Movement of the Welding Rods: A transverse or zigzag movement is given to the end of the rod in the molten metal.

The advantages of this method are:

(1) Variations in the width of the gap between the edges do not seriously affect the results.

(2) The technique is rapidly acquired, and small demand is made on the concentration of the welders.

(3) First-class results are continuously obtained.

(4) Distortion is reduced to a minimum owing to the constant cross-section of the welded zone.

(5) Improved economy due to the low consumption of gases as a result of the better utilisation of the available heat.

(6) Within the recommended range of thicknesses no chamfering of the edges is necessary.

The soundness and quality of the welds obtained by this method make it particularly valuable where the parts have to support considerable stress, as in the case of welded pressure vessels, containers, tanks, etc., where the two sides are accessible. The protecting effect of the reducing-zone of the flame is obtained, an advantage not secured by leftward or upward vertical welding, method "A."

Workmanship in Welding

With welded joints the allowable efficiency or ratio of joint-strength to the strength of the solid plate must depend upon a reasonably good standard of workmanship, since there can be 110 doubt that the quality of a welded joint is more easily affected by human inconsistencies than, say, a riveted joint. Defects which might be due to faulty workmanship are as follows:

(1) Imperfect fusion with parent-plate.

(2) Slag inclusions.

(3) Gas pockets.

(4) Contamination of weld-metal.

(5) Overheating of weld-metal or parent-plate.

Items (1), (2) and (3) and to some extent (4) will be detected by X-ray examination which must, in consequence, play an important part in the assessment of joint-efficiency. Item (4) indicates defects due to large fluctuations in arc-length, improper manipulation of electrodes, or, in the case of oxy-acetylene welding, an improper admixture of the gases in the flame. Under a careful system of supervision it is unlikely that these defects will be extensive throughout a welded seam, and the total effect of such local contamination (local not only in relation to the length of the seam but also in relation to the depth) need not be regarded with undue alarm. With regard to (5), serious overheating (that is to say, overheating which cannot be rectified by subsequent heat treatment) is unusual in the case of electric arc welding as carried out in high-class welding shops, because the factors of current values and electrode fluidity should always be subject to special control.

A possible cause of overheating, however, would be the maintenance of the arc in one position for too long a period. Fortunately, the defect would be readily revealed by visual inspection, due to the fact that it would be accompanied by serious undercutting at that particular position.

The purport of the foregoing remarks on workmanship is to show that this factor can, without difficulty, be accounted for in assessing the joint-efficiency, by providing:

(1) Adequate supervision of the welding work,

(2) Careful inspection of each run of weld-metal,

(3) X-ray examination of the finished joint.

It is the practice with reputable firms in Great Britain for an experienced supervisor to be responsible for the detailed inspection of each run of weld-metal as deposited in the welded joints of important pressure vessels. His duties also include the supervision of the arrangement of the vessel for welding, and the control of the current values.

Such close supervision must be regarded as an essential factor in fusion-welding, as it is only by means of this supervision that the reliability of workmanship can be accepted as equivalent to that involved in high-class riveting for boiler construction.

Material for Welded Pressure Vessels

In Great Britain, for steel pressure vessels the material usually specified is mild steel made by the acid or basic open-hearth process. The ultimate tensile strength of the plates lies between the limits of 26 and 30 tons per sq. in., or between 28 and 32 tons per sq. in., depending upon the tensile range specified. The minimum elongation required on a gauge-length of 8 ins. is 23 per cent, for the 26-30 ton steel and 20 per cent, in the case of 28 32 ton steel. In certain cases, the plates are normalised after rolling, and it is good practice to anneal (stress-relieve) the plates after bending them to the required shape.

Lower-tensile steel plates would be accompanied by lower carbon and manganese contents, but the range of silicon indicated in the above analysis permits the use of steel in either the fully " killed " or "semi-killed" condition.

It has long been the practice in riveted-boiler construction to use steel which may be said to be fully " killed," and which in consequence possesses a silicon content of about 0.2 per cent. For the purposes of welding, however, opinion turning in favour of low-silicon steel.

The requirements of other authorities in regard to plate-material for welding are as follows:

United States: The regulations here are based on the requirements of the Power Boiler Construction Code of the American Society of Mechanical Engineers. These requirements permit a range of steel for welding purposes; the maximum allowable carbon content for steel considered to be of weldable quality is 0.35 per cent.

Germany: In Germany regulations are in force in respect of forge welding of boiler-drums, but in general, fusion welding is only permitted in cases where the welded seam is covered by butt-straps. It is the practice, however, to give special permission, by means of Ministerial Decrees, to certain firms to enable them to manufacture fusion-welded boiler-drums without straps. The existing regulations are now being revised and a draft of new regulations has recently been submitted by the Committee appointed for that purpose.

The analysis of plate material most commonly used in Germany for welding purposes is:

Carbon...... from 0.1 to 0.3 per cent.

Manganese...... 0.5 per cent, (max.)

Phosphorus...... 0.05 „ „ „

Sulphur...... 0.05 „

Silicon...... up to 0.25 per cent.

Steel associated with this analysis is usually supplied in two ranges:

(i) From 34 to 41 kgs. per sq. mm., with an elongation of from 28 to 25 per cent, on 200 mm., and a reduction of area of from 80 to 70 per cent.

(ii) From 41 to 48 kgs. per sq. mm., with an elongation of from 25 to 20 per cent, on 200 mm., and a reduction of area of from 70 to 60 per cent.

Until recently steel of 48 kgs. per sq. mm. was considered to be the limit for good weldability.

Whilst the existing German regulations do not include any stipulations in regard to impact-testing of boiler-plate, the German Steam Boiler Committee have given considerable thought to the subject, and it is of interest to note that a Charpy test value of 10 metre-kgs. per sq. cm. is regarded as appropriate for plate thicknesses up to 15 mm. and 8 metre-kgs. per sq. cm. for plate thicknesses above 15 mm. These values are quoted in connection with plate material of the 34r-41 kgs. per sq. mm. quality, but in view of the well-known lack of consistency in impact-test results on rolled-steel plates the figures should not be regarded as a final criterion.

Recent progress in Germany has, however, made it possible to adopt welding also for steels of higher tensile strength, and for slightly alloyed steels. These steels include the above-mentioned qualities of mild steel with the addition of 0.5 per cent, of molybdenum, the steel having a tensile strength of from 35 to 50 kgs. per sq. mm.; ordinary mild steel containing 0.25 per cent, of molybdenum and 0.25 per cent, of copper, with a tensile strength of from 41 to 53 kgs. per sq. mm.; a special non-ageing carbon steel with a tensile strength of from 47 to 56 kgs. per sq. mm.; and this steel with the addition of 0.25 per cent, of molybdenum and 0.25 per cent, of copper. The above-mentioned steels are approved for electric-arc fusion welding, while in the case of water-gas welding, approval has been given for the use of a mild steel containing 0.25 per cent, of molybdenum and 0.25 per cent, of copper, with a tensile strength of from 44 to 56 kgs. per sq. mm., and for a steel having reduced ageing properties and a tensile strength of from 44 to 53 kgs. per sq. mm.

Switzerland: In Switzerland the regulations governing the construction of welded boiler-drums are issued by the Swiss Association of Owners of Steam Boilers (Schweizerischer Verein von Dampfkessel-Besitzern). These regulations, promulgated in January 1932, specify two qualities of mild-steel plate, and whilst both qualities may be used for welding.

Sweden: In Sweden, whilst considerable development has taken place in regard to the welding of structures and unfired pressure vessels, the welding of boiler-drums is not yet an accepted practice. No detailed regulations have been issued in respect of welded pressure vessels, but in general it can be said that the plate-material used for water-gas welding has a minimum tensile strength of 38 kgs. per sq. mm. and a carbon content of from 0.12 to 0.15 per cent.

The material is of the low-silicon type. For fusion welding, higher tensile strength is generally used, the material having a range of from 44 to 55 kgs. per sq. mm. and a carbon content of from 0.16 to 0.20 per cent.

Italy: In Italy, there are no detailed regulations except for a Ministerial Decree which deals broadly with the application of welding to pressure vessels, and which merely states that the plates are to be of a quality suitable for welding.

It is, perhaps, remarkable that although evidence indicates that Continental engineers are fully abreast of welding developments, no authoritative attempt has been made to establish a construction code governing the details of welding as applied to pressure vessels.

Even today the only authority to issue such a code in Europe is Lloyd's Register of Shipping, which, in July 1934, published " Tentative Requirements for Fusion Welded Pressure Vessels intended for Land Purposes."

Certain European authorities contemplate the acceptance of welded pressure vessels in their official regulations in that they do riot prohibit them. Others, such as the Swiss Association of Owners of Steam Boilers, go so far as to legislate for the design of welded pressure vessels, leaving the details of construction and testing to the discretion of the expert inspectors. In other cases, for example in Czechoslovakia, construction is generally carried out in accordance with the requirements of the Boiler Construction Code of the American Society of Mechanical Engineers.

Rules for Design of Welded Joints of Pressure Vessels

Joint Efficiencies: In considering the question "What is a reasonable figure to assign for welded joint-efficiency"? it would be well to take note of the joint-efficiencies and factors of safety allowed by those authorities which have legislated for welded joints in pressure vessels.

Great Britain: The only published rules governing design of Class I fusion-welded pressure, vessels are Lloyd's Register's "Tentative Requirements for Fusion Welded Pressure Vessels intended for Land Purposes." These are at present under revision, but in the meantime may be regarded as the accepted standard for both land and marine work in Great Britain.

United States of America: The Power Boiler Construction Code of the American Society of Mechanical Engineers permits a joint-efficiency of 90 per cent., combined with a factor of safety of 5. The formula for working pressure in lbs. per sq. in. is TS × *t x* E ~FS × R

where TS = the ultimate tensile strength of the plate in lbs./sq. in.

t = the minimum thickness of the shell-plate in ins.

E = the joint efficiency = 90 per cent.

FS — the factor of safety = 5.

R = the internal radius of the shell in ins. If the shell thickness is greater than 10 per cent, of the radius, the outer radius is to be used for R.

The specification governing the welding of pressure vessels issued by the Bureau of Engineering of the US Navy, states that the joint-efficiency of welded joints under the cognisance of this Bureau shall be taken as 80 per cent, of the strength of the parent-metal, except where the quality of the weld can be and is fully explored by radiographic or exographic photography, or other method satisfactory to the Bureau, in which case a joint-efficiency of 90 per cent, will be acceptable.

The joint "Code for the Design, Construction, Inspection, and Repair of Unfired Pressure Vessels for Petroleum Liquids and Gases", issued by the American Petroleum Institute in conjunction with the American Society of Mechanical Engineers Schuster (*Proc. Inst. Mech. Eng.*, 1930, 372) suggested Provisional Rules for Fusion-Welded Non-Fired Pressure Vessels, and gave the following recommendations about the method of fabricating pressure vessels:

1. *Scope of Rules:* The rules apply to vessels containing air or non-corrosive gas at a temperature not exceeding 600°F, and more especially when the diameter does not exceed 36 ins. For vessels with a diameter not exceeding

20 ins., the rules do not apply when the working pressure is 30 lbs. per sq. in. or less.

2. *Metfiod of Welding:* The rules are applicable to vessels welded by the metallic arc process with a flux-coated electrode or by the oxy-acetylene process.

   Note 1: The rule excludes carbon-are-welding, but in. special instances, where the design is approved, the carbon-arc process is permissible for the welding of flanges to tubes, and for the welding of dished ends to shell plates that are secured independently of the weld.

   Note, 2: It is recommended, especially when, the welding has been carried out by the carbon-arc or acetylene process, that fusion-welded parts should be normalised in a muffle furnace. The temperature should be maintained for a quarter of an hour after the article has been heated uniformly.

   As an alternative, though less effective, where serious damage from scaling or warping would result if it were normalised, the article should be heated uniformly to a temperature of 0-50°C, and then allowed to cool.

3. *Strength of Joint:* The rated strength of a welded joint shall at no time exceed half the strength determined by tests on reasonably well-made joints, made to similar form and subjected to similar stresses.

4. *Material:* The plate shall be of boiler-quality mild steel with a nominal breaking strength of 26 tons per sq. in. The strength shall lie between the limits of 24 to 28 tons per sq. in., but it shall be assumed that the strength of the plate when welded does not exceed 24 tons per sq. in.

5. *Plate Thickness:* The plate of a welded vessel exceeding 12 ins. in diameter.

6. *Factor of Safety:* The minimum factor of safety allowable on the rated strength of a welded joint shall be 4.

7. *Hydrostatic Test:*
   (a) The vessel shall be subjected to a hydrostatic test at not less than twice the working pressure, during which it shall be subjected to a hammer test; the pressure shall then be reduced by more than 50 per cent, of its value, after which it shall be again raised to the original value for 3 minutes.
   (b) The hammer test shall consist of the plate on both sides of all welds being struck sharp vibratory blows with a hammer not less than 2 lbs. in weight, the blows being struck 2 to 3 ins. apart and close up to the seam. The blows shall be applied as rapidly and heavily as possible, consistent with the metal not being indented or distorted.
   (c) When the factor of safety of the vessel exceeds 4, the test pressure shall be 50 per cent, of the bursting pressure, calculated on the rated strength of the welded joints.
8. *Stresses Imparted by Flexible Ends:* Where a joint or a portion of a joint, circular or longitudinal, is subjected to a bending action owing to the flexibility of an end, it shall be deemed that the strength of the joint is reduced by 10 per cent.

   This shall not apply to a longitudinal seam of which the ends are reinforced in an approved manner to resist bending stresses.
9. *Lap-Welded Joints:*
   (a) For a longitudinal seam, a lap weld with a single fillet shall at no time be permitted, and a lap weld with a double fillet shall only be permissible.
   (b) For a circular seam or for affixing an end plate when the shell plate is not constricted, i.e. turned in over the flange of the end plate, a lap-welded joint with a single fillet shall only be permitted.
   (c) When a lap-welded joint has a double fillet the

internal fillet shall be accessible for inspection. Where there is no ready means for inspection it shall be assumed that the fillet is non-existent, and accordingly for a circumferential seam the joint shall be rated as if it had a single fillet only and for a longitudinal scam it shall be prohibited.

10. *Fillet Welds:*

(a) With a fillet of a lap-welded joint the surface of contact in tension shall extend over the full thickness of the plate.

(b) A fillet of a lap-welded joint shall by preference have a convex profile. At no time shall the thickness at the throat be less than 0.69 times the plate thickness *t*.

Note: The throat thickness is defined as the minimum depth of the weld. Where the two contact surfaces of a fillet are equal, the rule requires that the contour shall show no appreciable concavity.

(c) With a fillet weld affixing a fitting to the wall of a vessel or a pipe, the throat shall have a minimum thickness of 0. 58Å, where *h* is the minimum height allowable for the surface of contact in shear.

Note 1: Where the two surfaces of contact of the fillet are equal, this throat thickness is attained by a fillet.

Note 2: The value required for the dimension *h* depends on the details of the design and manufacture of the joint.

11. *Butt-Welded Joints:*

(a) When the diameter of the vessel does not exceed 24 ins., the strength of a butt-joint with the plate welded at only one side shall be deemed to be reduced to half the strength of a butt-joint with the plate welded at both sides. For larger diameters this form of construction shall at no time be permitted.

(b) When the inside surface of a butt-joint is not accessible for inspection, it shall be assumed that the plate is welded at one side only.

(c) The edges of a plate jointed by a butt-weld shall be bevelled, and the included angle of the weld-metal shall be not less than 90° for either a single or a double V weld.

12. *Design and Attachment of Dished Kinds:*

(a) A dished end plate shall have an inside radius not greater than the internal diameter of the shell. The thickness of metal shall be such that the working pressure or the allowable stress to comply with rule 13 is not exceeded, but in no instance shall it be less than the thickness of the shell, and for vessels exceeding 12 ins. in diameter the minimum thickness shall be 1 in.

(b) When a dished end is flanged, the length of the flange shall be not less than 1 ins. for an end convex to the pressure, nor less than 2 ins. with the end concave. The minimum length of the flange shall at no time be less than four times the thickness of the plate. An increase above these specified minimum values is desirable.

Copper Steam-pressure Vessel

The copper used in the construction of vessels should be in the hard condition, not annealed, and when forming a cylindrical shell subject to internal pressure, or when in tension not exposed to heating by furnace or flue gases, a maximum tensile strength of 5000 lbs. per sq. in. may be allowed. Superheated steam should not be used in vessels made of copper.

When copper is in the annealed condition, as a rule the maximum tensile stress should not exceed 3000 lbs. per sq. in.

The two most common forms of joint used in connection with copper plates are the brazed and the welded joint.

Occasionally joints are clamped and brazed, riveted and brazed or riveted and soldered.

For the purpose of calculation the efficiency of an ordinary brazed or clamped and brazed joint should not be considered higher than equal to 50 per cent of the plate strength.

The design of riveted joints should be carried out on the same lines as for steel plates with steel rivets, but in the design of the copper joint, account requires to be taken of the effect on the design of the lower strength of copper.

Particularly, it is to be noted that the grip of the copper rivets holding the joint together will not be so great as in the case of steel rivets; consequently, on these grounds, a closer pitch will be desirable than in the case of steel rivets. Also, the copper rivet being weaker than a steel rivet of the same side, it will be necessary in a given pitch for the copper rivet to be of somewhat larger diameter than a steel rivet would be for the same pitch. The effect of this will be to modify the percentage strength of joint which would otherwise be obtainable.

Protection of Steam-heated Pressure Vessels against Overpressure

It is essential that steam-heated pressure vessels connected to steam boiler plant should be protected against dangerous overpressure if they are designed to work at a pressure lower than that of the boiler plant. This requirement is stressed in the recent Factories Act 1937, and may be illustrated by the following example:

> Assume that a reducing valve having an inlet bore of 1 in. diameter is supplied with steam at 120 lbs. pressure per sq. in. and is set to reduce the pressure in a low-pressure vessel to 10 lbs. per sq. in., and that no steam is being condensed in the vessel or discharged except through the safety valves. Also assume that the reducing valve has become deranged and stuck fast in the full open position, thereby allowing high-pressure steam to enter the

low-pressure main. The problem to be solved is what should be the safety-valve area under such conditions to ensure that the pressure in the low-pressure vessel is not increased above 10 lbs. per sq. in.

The problem can best be studied from the laws governing the flow of steam through orifices, the supply pipe being considered as an inlet orifice to the low-pressure vessel and the safety valves as outlet orifices therefrom.

The Hollow Forging of Pressure Vessels for Very High Pressures.

The increasing demands of recent years for large high-pressure boiler-drums and chemical reaction vessels has led to the development of hollow forging of such vessels because the wall thicknesses are too great to permit of riveted construction, and in the case of chemical vessels, the materials are frequently of such a composition that welded construction is inadvisable.

Such hollow-forged vessels may frequently have a finished weight of 60-70 tons, which necessitates an ingot from which to forge the vessel of 160-180 tons, and this in turn has led to marked developments in ingot moulding.

In the manufacture of hollow forgings until recently the usual practice was to an neal the ingot after stripping it whilst still hot from the mould, afterwards allowing it to cool very slowly. When cooled, a hole was trepanned through the centre, the object of which was to produce a core of metal which could be closely examined, and also to remove the less pure material from the axis of the ingot.

The next operation was to enlarge the hole by an hydraulic pressing operation after the ingot has been reheated to the necessary forging temperature, and this produces a long tubular forging. Subsequently the drums are normalised, the outside is rough machined and the interior bored. After reheating the ends are closed for manhole openings, in the case of boiler-drums, or are reduced and then flanged in the case of chemical reaction vessels.

It will be realised that the above process is a lengthy and costly one, and attempts have recently been made to reduce both the labour costs and the production time. The first of these consists in hot piercing or punching whereby a hole is made through the ingot without cooling it. In this process, the feeder head at the top and the surplus metal at the bottom of an ingot is removed by a large cutter, driven through the ingot by means of an hydraulic press. The remaining body of the ingot is then placed vertically beneath a press, and a tubular punch driven axially through, thus removing a core from the centre of the ingot.

Another method recently adopted involves the casting of a hollow ingot, and it has several great advantages. Firstly, economy in raw material is secured, and secondly considerable reduction in time of manufacture is obtained.

Until recently, large pressure vessels were available only in the carbon steels, but now large ingots can be obtained either by open hearth or electric furnace process in alloys such as 3 per cent, and 6 per cent, chrome molybdenum, and the 25 per cent, nickel-chrome molybdenum types. The subsequent heat treatment of such large masses of alloy steel is a difficult and costly process, but in view of the special features of such alloys, manufacturing costs are not prohibitive.

Steam-pipes

It is used in nearly every branch of chemical engineering, and it is of importance., therefore, that strum supply pipework be well designed, adequat e in size or the purpose desired, well covered to prevent unnecessary heat losses, properly drained to prevent possibilities of damage by water hammer and to provide dry steam at the point of use.

With the introduction of mild steel pipework at the beginning of the present century, the manufacture of cast-iron and wrought-iron steam-pipes steadily declined, and within the limits of susceptibility of this material practically all steam-pipe installations are now of mild steel throughout. For extreme temperatures at high pressure alloy steels are imperative, but since these conditions are of recent growth the technique is by no means stabilised.

J. A. Alton gave an interesting paper on modern high pressure-high temperature pipework, a summary of which is as follows:

> Pipework for severe operating conditions must he considered from three main aspects: (a) the raw material of the pipe, (b) the method of manufacture, and (c) the working conditions, both as regards pressure and temperature of the gas or fluid being conveyed, and the arrangement of the pipe.

At the higher temperature ranges, the "creep" of steel must be taken into consideration; and if the maximum life of the pipework be taken as twenty years, then it is reasonable to assume that the actual time which the plant will be under service condition is, say, 100,000 hours.

Since the permissible creep in this time may be taken as 1 per cent., it follows that the allowable creep per inch per hour is one-tenth of one-millionth, i.e. 10^{-7} ins. Taking 60 per cent of this creep as the maximum permitted to allow for contingencies, the following stresses are suitable for 0.17 per cent, carbon steel.

In using the foregoing formula and data in determining pipe dimensions it must be borne in mind that manufacturing difficulties do not permit of 0.15 per cent, carbon steel tubes with wall thicknesses necessary at pressures approaching, say, 1400 lbs. per sq.in. at the maximum working temperature given, viz. 900°F.

Recourse must then be had to alloy steels, and since such steel for pipe fabrication has to be capable of being worked at high temperatures in the manufacturing process without subsequent elaborate heat treatment, and of being welded for flanges, etc. without detriment to its mechanical properties in the process, great care is necessary in selection.

Steam-pipe Connections and Flanges

To connect saturated steam pipework at pressures not exceeding 150 lbs. per sq. in. and pipe diameter of 2 ins., it is satisfactory to use standard screwed bronze unions, screwed tees, bends or sockets, but above those pressures and diameters it is advisable to use flanges, and flanged tees or bends.

The dimensions of flanges and flanged fittings, and the method of securing flanges to the tube, depends largely upon the working pressure and temperature. To avoid carrying an excessive variety of flanges it is advisable to standardise on as few different sizes as the plant will permit.

Flanges may be fixed to pipes by welding, screwing, expanding or riveting, or by a combination of two or more of these methods. Expanding alone is definitely not recommended for even the lowest pressures.

Welded-on flanges can be supplied for all sizes, temperatures and pressures. The riveted flange is not recommended for high temperature and it is difficult to fit with sizes less than 7 ins. diameter.

Screwed flanges are suitable for pressures up to 200 lbs. per sq. in., and screwed and expanded flanges for pressures up to 350 lbs. per sq. in. at temperatures not exceeding 650°F.

Regarding cost, the riveted flange is the most expensive, and the screwed-on flange the cheapest type.

In the case of screwed- and expanded-on flanges it is usual for the expanding to be done with proper appliances at the tube-maker's works.

It is important in such cases to note that the pipe is screwed with a vanishing thread, so that the pipe is not unduly weakened by the expanding process at the end of the thread. If such pipes are likely to be subjected to torsional stress it is well to reinforce the attachment by a light weld fillet at the back of the flange.

Such flanges should be welded to the pipe which is a type

which can be well recommended. The chief features of this welded connection are:

1. The main weld is a butt-weld and is consequently either in tension or compression but never shear, and consequently a minimum of weld material is required.
2. The bending stresses on the butt-weld arc reduced to a minimum.
3. In subsequent facing of the flange none of the weld-metal is removed.

The butt-weld should be by the metallic; arc process, and the fillet either by metallic arc or carbon arc process.

When the steam pressure exceeds 350 lbs. per sq. in. and the temperature 750°F, great care must be exercised in choosing the type of flanged connection from pipe to pipe, and pipe to fittings or valves.

The design of pipe flanges for very high-pressure steam lines has naturally received the attention of the British Standards Institution, who in 1932 issued a tentative specification for flanges of pipework and fittings to be used for working pressures from 900 up to 1400 lbs. per sq. in. at temperatures up to 800°F, but extended by a reduction in the working pressure to 900 lbs. per sq. in. to a maximum of 900°F. In issuing this specification the Committee pointed out that the recommendations were in advance of practice and knowledge and the tabulated data were thereupon issued as a *tentative* standard only.

For the very highest pressures there are three forms of welded high-pressure pipe connection which have proved successful in practice. The first was developed in America and is known as the "Sarlun" joint. The Sarlun joint is a development of the "Sargol" joint, which was in effect the old well-known "Van Stone" joint. In it the collars are the same as in the Sargol joint, but the collar is turned down circumferentially, thus leaving on the face side of the collar a raised V-shaped circumferential ring. This being thin, requires much less heat in the welding operation. The faces of the collar are scraped and made steam-tight before welding.

The second form of seal-weld joint is the " Corwel", developed in Great Britain. This joint uses a loose flange of similar shape to the " Sarlun," but the collar is produced by raising a corrugation on the end of the pipe and rolling this back, thus producing a double or folded collar. The face is machined smooth, and a heavy weld is used without difficulty. The bolts used for the Corwel joint are nickel-chrome molybdenum steel and are not formed with head but are screwed fine thread right through.

High-temperature Steam Experience at Detroit

The authors report experiments on two high-pressure high-temperature steam installation, one at the Trenton Channel power-station and one at the Delray power-house No. 3. The first experimental plant at Trenton started up in 1929 comprised a separate oil-fired superheater and piping system designed to operate with steam at 1000°F, and later at 1100°F, and this was operated for two years to give data for the second stage of the experimental work which comprised the installation of a 10,000 kW. steam turbine with necessary equipment for working at 1000°F. The working pressure throughout was limited to pressures not higher than 400 lbs. per sq. in., and it was considered that the higher temperature, without the added complication of very high pressure, would provide the information most desired.

Description of Equipment: The Trenton equipment comprised an oil-fired superheater supplied with 400 lbs. 700°F. steam from the power-station main header. The superheater was of the radiant type, and was capable of adding 400°F of superheat to 6000 lbs. of steam per hr. from the initial temperature of 700°F. The high-temperature piping system consisted of a valve, eight flanged joints, and a de-superheater in order that the steam used in the experiments could subsequently be used in the house turbine at 700°F.

The superheater was made of low-carbon steel tubing in the section where the steam temperature did not exceed 800°F, and the remainder of the superheater and the high-pressure piping system were made from stainless steel with analysis C 0.06-0.09, Cr 17-20, Ni 7.10, Si 0.5. The steam piping is 5′ diameter, outside with thick walls. The loose companion flanges, valve and desuperheater castings were made of similar material to the above, except that the carbon was increased to 0.25 and the silicon to 2.0-2.5 per cent.

The flange bolts were made from chrome-tungsten vanadium steel. The pipe ends were provided with Van Stone Sarlun laps, the faces of which were provided with a serrated finish to provide data for gasketed as well as seal-weld joints. All piping joints were constructed to the 600 lbs. A.S.A. standard. The original line has been somewhat modified by the addition of experimental joints of heavier construction.

Numerous gasket materials were tried, and the best was found to be 3-in. Monel metal in plain sheet form.

Several different classes of low-alloy stools were installed for experimental purposes in both Trenton and "Delray installations. Apart from savings in material cost, such low alloys are more readily fabricated. Among the castings tried were 46 per cent. Or 1 per cent. W 0.6 per cent. Mo; a 5′ outside diameter length of 4.6 per cent. Cr 1 per cent. W; a valve body of Hadlields Era 131; medium- and high-carbon calorised stool together with a nickel-chromium molybdenum steel adopted by the turbine manufacturers for all parts of the turbine subjected to high-pressure steam.

A few tubes of medium- and high-carbon calorised steel were installed in the Trenton superheater, arid 0-35 C steel tubes coated with aluminium were installed in the middle section of the Delray superheater, together with a like number of coated tubes of 18.8 and 12.14 per cent, chrome-nickel austonitic steels. Metal gaskets were used for all unwelded joints, and the flanges were made of low-chrome nickel stool.

Results of both physical and microscopic examinations made on various materials after service from the two installations have shown slight change after service under the experimental conditions, but in no case have the changes been sufficiently serious as to warrant replacement of affected parts. Six different classes of alloys were examined.

Three different sections of 18-8 material removed from the Trenton installation showed evidence of carbide segregation at the grain boundaries; and each successive sample revealed widening of these boundaries. The physical properties appeared to indicate slow embrittlement, but such changes were not considered to cause concern as to their continued service. The authors recommend the development of special alloys to avoid the chromium-carbide segregation.

The calorised superheater tubes and headers of medium- and high-carbon steel were removed from the Trenton superheater after 3821 hours' service, of which 3305 hrs. were above 1000°F, and stresses due to all circumstances as high as 3000 lbs. per sq. in. Approximately 10 per cent, of the calorised surface coating had disappeared and the remaining 90 per cent, was quite brittle. The inside of the tubes was coated with 0.04-0.08 thickness of magnetic oxide, and measurements showed creep rates of 2 per cent, per 1000 hours' service.

The Dolray calorised and sprayed superheater tubes were not removed for inspection, but the external surface was apparently in quite good condition.

A valve at the outlet of the Trenton superheater made from 0.4 per cent. Or 1 per cent. W was removed for examination after 12,995 hrs., 10,818 of which were at a temperature of 1100°F The valve was 600 lhs. A.S.A, construction, and the test results show that the material of the valve body did not undergo embrittlement, but that the long service at high temperature tended to increase the toughness and ductility. Experience with Nitralloy showed that it is unsuitable for valve parts.

Samples of the nickel-chrome molybdenum steel from the

throttle valve of the turbine after 3800 hrs., 1450 of which were at 1000°F, showed but little change in physical properties.

Various bolting materials were tried with varying results. Bolts made from this material were heat-treated by oil quench from 1650°F followed by 1100°F tempering, and the threads were cut after heat treatment to prevent the formation of cooling cracking the thread roots.

Pipe Joint Experience

Many and varied troubles were experienced with pipe jointing, but from experience gained it was found possible to construct two joints which successfully withstood 16,000 hrs. at steam temperatures of 1000°F. The authors state that full weld-joints appear to be the answer to the high-temperature problem.

The successful 5-in. bolted flange joint was of the Van Stone type with serrated flanges of KA2 material. The bolt stress was limited to 10,000 lbs. per sq. in., and one joint of this type had flanges as 900 lbs.

No material dimensional changes were noted in the 8-in. Delray pipe stressed to 6400 lbs. per sq. in., or in any of the fittings, and measurement of a section of 51 in. outside diameter chrome-tungsten tube after service in Trenton likewise showed no change after 7468 service hours.

In a concluding paragraph, the authors state that the use of steam equipment up to 1000°F is entirely feasible, but this is qualified by the statement that if such operation demands existing high-priced alloys, it is scarcely workable. Further metallurgical research will no doubt, however, produce durable materials at a much lower cost, when such high temperatures will be justifiable.

Bends, Expansion Bends and Joints

Bends: In designing large bends for steam mains three rules should be observed: (a) all bends should be formed on long lengths of tube wherever possible, (b) all bends should have at each end a straight portion not less than one and a half pipe diameters, and (c) all bends should be of as large radius as possible. In some installations, however, the layout may necessitate the use of bends of short radii, and in such cases creased or corrugated bends may be used.

It must be realised that creased and corrugated bends cause largely increased pressure drop and their use is not recommended unless the layout renders this imperative. Where bends of still shorter radii are necessary, these can be formed by guesting, i.e. wedge-shaped pieces are cut from that side of the tube which is to form the inner curve of the bend and the edges of the opening are bevelled. The tube is then bent, forcing the cut edges together, and these are then welded. As an alternative to such gusseted bends cast steel elbows may be employed.

With small radii and very thick bends, the bending process may thin the outside wall of the bend by 30 per cent. Creasing the inner wall makes it possible to bend pipes of the dimensions given to a smaller radius without thinning the outer wall, and such creasing definitely increases the flexibility of the pipe. Creasing a bend enables it to be made in large diameters.

The fully corrugated pipe can be made to smaller radii even than the creased pipe. They are more flexible. No case of corrosion in the walls of the corrugation has come to the notice of Mr. Aiton, who has had very wide experience of this class of work.

Expansion Sends: Provision for expansion must be made in the design of every steam main, and this is best done by arranging a sufficient number of ordinary bends to allow for all movement. Where it is not possible to do this, expansion bends of this type may be used. In arranging the pipe movements, care should be taken that branch connections to the main do not cause restriction of such movement, otherwise serious damage to the main may result.

Sliding expansion joints of the stuffing-box type are used

where space does not permit of the use of expansion bends, but such joints should only be used for low pressures and temperatures.

Great care is necessary in erecting and packing such joints, as leakage from the gland will otherwise result. Copper expansion bellows are sometimes used, but they are not recommended for pressures higher than 10-20 lbs. per sq. in., or where torsion or excessive movement may possibly arise.

Erection of Expansion Bends

To use an expansion bend to the best advantage it is necessary to spring it during erection. In doing this, care must be taken to prevent any movement of the main at the anchors, since this would mean the loss of so much of the cold springing and consequently an increase in the stress imposed on the bend in the working condition if it is still to take up the same amount of expansion movement.

Pipe anchors in suitable positions are essential for the satisfactory working of all types of expansion bends or expansion joints, and even where these are not employed anchors are frequently necessary to regulate the movement of the main. In long mains, guides to maintain the alignment are necessary both for expansion joints and expansion bends—for expansion joints because any misalignment will probably cause the joint to jam; for expansion bends because the end load necessary to produce the safe travel of the bend may be greater than the adjacent pipes can supply without themselves bending.

Expansion of Steam Pipes

Formula for calculating Expansion: The amount of expansion which will occur on any steam main can be calculated from the formula:

$E = KLf$

where E = amount of expansion, in ins.

L = length of piping under consideration, in ft.

t = maximum temperature variation in °F,

and K lias one of the following values, depending on the range of temperature.

The total variation in temperature should be determined basing on a minimum temperature appropriate to the local conditions.

Branches and Connections to Mains

For low pressures, i.e. up to 150 lbs. per sq. in., and for pipes below 5 ins. diameter, screwed tees are quite suitable for branch connections. For higher pressures and larger pipes it is usual to take branch connections from flanged tees or from riveted or welded-in branches on the main.

In attaching welded branches most makers cut out the opening in the main or manifold by oxy-acetylene cutter and then shape the contour of the branch piece to the pipe opening. The connection should then be welded with a good fillet. When possible branches should be welded on at right angles to the main, as the large flat surfaces formed at the junction of an angle branch with the main are subject to a much higher stress than the cylindrical parts. Bosses welded on are often used in place of branches for small connections to mains. They are either tapped to receive the screwed end of a pipe or tapped for studs for flange connection.

If desired and over 10 ins. diameter riveted branches can be used where the pressure and temperature are low. Such branches should be placed near the ends of the main to permit of efficient holding of the rivet-heads while riveting. The riveted branches should be of pressed steel of similar design to boiler nozzles or stand-pipes, and should be of thicker gauge than the standard pipe size so as to allow for thinning in flanges, and give sufficient landing edge for caulking.

Steam Separators and Receivers

Complication in separator design is not desirable; it usually adds little to the efficiency and often causes unnecessary loss of pressure.

Steam separators depend for their powers of extracting water from steam on the following principles:

(1) Sudden alteration of direction of flow.

(2) Reduction in velocity of flow.

Separators are designed to utilise both of these principles. Separators should have hand holes in the body where the drain brandies are not sufficiently large to take their place.

Steam Receivers: Riveted steam receivers are suitable for steam pressures up to 500 lbs. per sq. in. working pressure at temperatures not over 650°F The receiver shell is formed from one plate, the longitudinal joint being butted, fitted with double butt-straps, and electro-welded at each end to ensure steam tightness. Each receiver end is made from a single plate pressed to shape in one heat and machined to ensure a good fit in the shell, which is bored out to receive the ends. The ends are of special form, giving the maximum strength combined with a flat large enough to accommodate the usual elliptical manhole for access to the interior. Stand-pipes for this type of receiver are solid forged, and double riveted to the receiver.

For working pressures above 500 lbs. per sq. in. at temperatures exceeding 650°F solid forged receivers are recommended. On these the stand-pipes are solid forged and are welded to the receiver.

Steam-piping Supports

The supporting of steam-pipes is closely connected with the provision for movement of expansion and contraction, but the supports proper should, as far as possible, permit free movement of the pipes in any direction, and any necessary constraint should be provided by anchors.

Where pipes are subject to vertical movement or to vibration, spring supports should be employed to avoid excessive stresses which might be set up in the pipes by the use of too rigid supports. In designing such supports the selection of springs of the proper strength is of importance. If the springs are too weak they will allow undue vibration to take place, whilst if too rigid they will not relieve the stresses in the pipeline. The spacing of supports is largely determined by the position of floor girders, joints, columns or other structural features, but supports must not be placed too far apart. As far as possible, pipe joints should be placed close to the points of support.

Although in many cases pipelines can be allowed free movement, subject only to constraints imposed by the fixing of their end flanges to the units which they connect, it is sometimes necessary to introduce fixed anchors to minimise the stresses which might otherwise be set up at certain points of the system. The location of anchors must be very carefully considered in relation to the system as a whole.

As auxiliaries to fixed anchors, limiting or guiding anchors are sometimes required to prevent movement in some particular direction or to maintain the correct alignment of the pipes under the loads imposed by expansion.

Drainage of Steam Mains

The proper drainage of steam pipework and receivers is of the utmost importance, and such drainage should include provision for the immediate removal of the water formed by condensation, and for the water which accumulates in the pipes after steam is shut off the main.

It is to be noted that water will not readily flow against the direction of steam flow even if the pipes are given a steep inclination. A normal steam speed of, say, 100 ft. per sec. is comparable to a wind velocity of 60 m.p.h., which is likely to cause water to back up against the direction of flow of the pipes.

Where possible, the boiler junction valves should be the highest point of the range, otherwise the drainage of a branch steam-pipe to a boiler which is out of action is difficult and the presence of water in this pipe is particularly liable to set up water hammer. If necessary and possible, the boiler junction valve may be raised by inserting a stand-pipe between the boiler and the valve.

To unsure that pipes can be cleared of water before steam is admitted, it is necessary that there be a drain and an efficient steam trap at the lowest part of each section that there be also means for admitting air to each section, preferably at the highest point.

All drain connections should be of ample bore to prevent choking, and should be made in the mains at such points as at valves of smaller diameter than the main, expansion bends in vertical planes, the foot of each rising bend and all similar positions. It is specially important that branch pipes rising from the main and provided with stop valves next the main should each have a drain connection to prevent water from accumulating when the branch pipe is not in use. The occurrence of such danger-points can often be avoided by slight changes in the design.

To take drain connections from mains either bosses welded to mains, drilled and tapped to take the screwed end of the drain pipe, should be used, or branches welded on and flanged to match the flange on the drainpipe. If more water is expected than could be satisfactorily discharged by a drainpipe connected to the main, a drain pocket of proper size welded to the main, flanged and fitted with a blank flange or with the end closed, should be employed, the blank flange or closed end being fitted to take the drain connection.

Steam Traps

There are many designs of steam traps on the market, and care should be taken to select the proper type for a particular purpose. Traps may be classified into two main types, viz. lifting and non-lifting. The latter is used when the condensate is discharged at a level lower than the trap and where the discharge of the condensate imposes no back pressure upon the trap. Lifting traps are used to raise the condensate to a level higher than the trap, in which case the trap must frequently be designed for a considerable back pressure.

Every trap should be provided with some means of testing its condition while at work, and with a by-pass to permit of its being removed for repairs.

Many manufacturers specialise in the manufacture of all types of steam traps, and when purchasing such appliances the fullest information should be given in the specification and enquiry in order that the most suitable type is supplied.

Steam Reducing Valvex

Steam at reduced pressure is frequently required in many departments of a chemical works. It is usual to supply steam at high pressure to the department concerned, and reduce, close to the point of use, thereby keeping the steam supply pipe as small as possible.

The principle of most reducing valves is the balancing of a double beat valve by assisting the low-pressure side of the valve by a spring, which can be varied in strength for various reduced pressures. In this way the pressure in the reduced main is automatically controlled, since if the pressure in the reduced side rises beyond that arranged for, the double beat valve closes and steam is not again admitted until the reduced pressure falls.

Relay reducing valves are desirable especially with widely fluctuating loads to prevent hunting of the valve arid wide variations in steam pressure.

Flow of Steam Pipes

In determining the size of a steam main the known factors are usually the quantity of steam to be conveyed, the initial pressure and temperature, and the length of the main. These are not sufficient definitely to fix the size of the pipe, as the permissible pressure drop or the velocity of flow must first be decided upon. These two factors are interdependent, and both are subject to limits which it is not wise to exceed.

It is not possible to lay down any hard and fast rules on the subject of velocities of steam flow, as so much depends on the circumstances of each individual case. In addition to the velocities being limited by the permissible drop in pressure, there are certain maximum values which should not be exceeded. One of the chief limiting factors is the erosive action of the steam on the valve seats and other similar exposed parts. This action is more pronounced with wet than with superheated steam. The velocity of wet steam should, therefore, not be as high as that of superheated steam.

Examples

An outstanding example of a large industrial plant necessitating the transmission of steam over long distances is that installed at the Billingham synthetic ammonia and nitrates works of Imperial Chemical Industries. Briefly, this plant consists of eight three-drum pulverised-fuel boilers, each designed to operate normally at an output of 215,000 lbs. of steam per hour. The normal saturated steam pressure is 715 lbs. per sq. in. abs., but the supply pressure to the high-pressure distributing receiver is 675 lb. per sq. in. abs., the maximum steam temperature being 458°C.; and the steam is supplied to three 12,500 kW. back-pressure turbines exhausting at 290 lbs. per sq. in. abs. and 345°C into a low-pressure receiver from which it is passed to (a) process plant steam mains and (b) two 12,500 kW. condensing turbo-alternator sets. From the latter the steam is wholly extracted and is used in four feed-water heaters and in an unusually large quadruple-effect distillation plant which supplies make-up feed water to the boilers. The low-pressure receiver is equipped to act as a desuperheater for any steam passed directly to it through a reducing valve from the high-pressure line, There are altogether seven feed heaters, and the feed temperature at inlet to the Foster steaming economisers is 205°C.

The total steam output of the boilers is 11,900,000 kg. per day, of which 6,930,000 kg., or about 57 per cent; is used for the process plants. About 42 per cent, of this amount—i.e. about 24 per cent, of the total output—cannot be recovered, hence the large capacity of the distillation plant installed. The maximum distance of transmission of steam in this plant is stated to be 1 mile, both for the high-pressure (290 lbs. per sq. in.) steam and for a 30 lb. per sq. in. low-pressure supply. The 290 lbs. per sq. in. process plant steam is utilised in driving non-condensing reciprocating engines and turbines, the exhaust from which is used for heating purposes in evaporation vats. A number of mixed-pressure turbines serve to maintain a balance between power and heating steam requirements.

An American plant of considerable interest from the point of view of process steam supply is the deepwater steam station erected on the east bank of the Delaware river in New Jersey (*Power Plant Engineering*, 1929). This station is run in the joint interests of three large companies, each of which has its own plant in the central building. The first two—which are power supply companies—have cross-compound turbo-alternator condensing units, of 53,600 kW. capacity, supplied from a pair of Babcock and Wilcox boilers (one standard and one reheat), with steam at 1215 lbs. per sq. in. abs. and a temperature of 385°C.

The third company the Du Pont do Nemours Company—has a high-pressure turbine of 12,500 kW. capacity, identical in size and construction with the high-pressure turbines of the two larger sets. It is also supplied with steam at 1215 lbs. per sq. in. abs. and 385°C from two standard Babcock and Wilcox boilers, and the whole of the power generated is delivered to the Du Pont Company. At normal load, this turbine exhausts 530,000 lbs. of steam per hour at 400 lbs. per sq. in. abs. into seven single-effect high-pressure evaporators, which in turn provide hourly 400,000 lbs. of steam at 180 lbs. per sq. in. abs. from raw water. This medium-pressure steam is superheated to 227°C by live-steam reheaters and delivered by two 16-in. mains, 1500 ft. long, to the works of the Du Pont Company. The exhaust steam from the turbine is completely condensed in the, evaporators, and returned directly by centrifugal pumps to the suction of the boiler feed pumps.

The object of this rather unusual arrangement is to avoid the necessity of pumping back from the Du Pont works any condensate that might be available from the process plants. This has the disadvantage of requiring the use of evaporator plant of large capacity, arid of reducing considerably the available "Bankine" heat drop of the steam through the reduction of pressure.

The latter point is not, however, of great importance if the steam is used wholly for heating purposes. The Du Pont plant is interconnected with the systems of the two other companies. Should the process steam requirements necessitate the turbine working at full load, the excess power generated is supplied to the other electrical systems; on the other hand, if the steam requirements are low and the electrical power developed in the turbine in consequence insufficient, the deficiency can be made up from the other power systems.

Pipes and Pipework for other than Steam Services.

The chemical engineer today has a wide choice of materials for tubes for handling liquids, gases, etc., used in chemical process work. This firm also manufactures steel capillary tubes. Where thermal conductivity is a requirement these capillary tubes can be obtained with a copper tube drawn tightly over the outside. Acdc & Pollock also make tubing in a wide variety of special steels for use in the hydrogenation of coal and oil, the synthesis of ammonia, etc.

Most of the non-ferrous materials suitable for chemical plant are now available in tube form, e.g. copper and copper alloys, aluminium, nickel, Monel metal, silver, lead, etc. In addition, chemical stoneware, hard rubber, fused silioia, and other non-metallic materials including plastics (e.g. Bakelite, Haveg, etc.) are now made into tubes.

In the latter connection the following description of a new plastic will be of interest. This material is based on poly vinyl chloride, and tubes are now made under the trade name " Mipolam".

It is absolutely resistant to vegetable and mineral oils and alcohol, (it shows but a slight increase (up to 0.5 per cent.) in weight after 6 months' immersion in petrol), acetic acid, sodium carbonate, caustic soda (up to 50 per cent.), nitric acid 30 per cent., hydrochloric acid (30 per cent.), sulphuric acid including concentrated acid at 45°C. It is unsuitable for use with acetone, ether, benzene, cyclohexamone, ethyl acetate, chlorohydrocarbons and nitrating acids. It is non-inflammable.

Tubes in this material can be sawn, filed, turned and drilled with ordinary metal working tools. Since the material becomes plastic above 80°C. it can be softened at any convenient source of heat, and then moulded as desired.

To join successive lengths of Milopam one end of one tube is warmed slightly and then expanded to receive the end of the next tube. The joint is made by a special adhesive. T-pieces, etc., can be made in a similar way.

Insofar as fabrication of chemical pipework is concerned, in view of the wide variety of materials of construction, and the special nature of most chemical process pipework, it is impossible to give any detailed rules. In regard to large pipework for gases and liquids, the principles given for steam pipework are often applicable.

4

Chemical Processes

In a "Process (science)" sense, a chemical process is a method or means of somehow changing one or more chemicals or chemical compounds. Such a chemical process can occur by itself or be caused by somebody. Such a chemical process commonly involves a chemical reaction of some sort.

In a "Process (engineering)" sense, a chemical process is a method intended to be used in manufacturing or on an industrial scale to change the composition of chemical(s) or material(s), usually using technology similar or related to that used in chemical plants or the chemical industry.

Neither of these definitions is exact in the sense that one can always tell definitively what is a chemical process and what is not; they are practical definitions. There is also significant overlap in these two definition variations. Because of the inexactness of the definition, chemists and other scientists use the term "chemical process" only in a general sense or in the engineering sense. However, in the "process (engineering)" sense, the term "chemical process" is used extensively. The rest of the article will cover the engineering type of chemical process.

Although this type of chemical process may sometimes involve only one step, often multiple steps, referred to as unit operations, are involved. In a plant, each of the unit operations commonly occur in individual vessels or sections of the plant called units. Often, one or more chemical reactions are involved, but other ways of changing chemical (or material) composition may be used, such as mixing or separation processes. The process steps may be sequential in time or sequential in space along a stream of flowing or moving material.

For a given amount of a feed (input) material or product (output) material, an expected amount of material can be determined at key steps in the process from empirical data and material balance calculations. These amounts can be scaled up or down to suit the desired capacity or operation of a particular chemical plant built for such a process. More than one chemical plant may use the same chemical process, each plant perhaps at differently scaled capacities.

Such chemical processes can be illustrated generally as block flow diagrams or in more detail as process flow diagrams. Block flow diagrams show the units as blocks and the streams flowing between them as connecting lines with arrowheads to show direction of flow.

In addition to chemical plants for producing chemicals, chemical processes with similar technology and equipment are also used in oil refining and other refineries, natural gas processing, polymer and pharmaceutical manufacturing, and water and wastewater treatment.

Alchemical Processes

Solvent: A solvent is a liquid that dissolves a solid, liquid, or gaseous solute, resulting in a solution. The most common solvent in everyday life is water. Most other commonly-used solvents are organic (carbon-containing) chemicals. These are called organic solvents. Solvents usually have a low boiling point and evaporate easily or can be removed by distillation, thereby leaving the dissolved substance behind. Solvents should therefore not react chemically with the dissolved compounds—they have to be inert.

Solvents can also be used to extract soluble compounds from a mixture, the most common example is the brewing of coffee or tea with hot water. Solvents are usually clear and colourless liquids and many have a characteristic odour. The concentration of a solution is the amount of compound that is dissolved in a certain volume of solvent. The solubility is the maximal amount of compound that is soluble in a certain volume of solvent at a specified temperature.

Common uses for organic solvents are in dry cleaning (e.g. tetrachloroethylene), as paint thinners (e.g. toluene, turpentine), as nail polish removers and glue solvents (acetone, methyl acetate, ethyl acetate), in spot removers (e.g. hexane, petrol ether), in detergents (citrus terpenes), in perfumes (ethanol), and in chemical syntheses. The use of inorganic solvents is typically limited to research chemistry and some technological processes.

Polarity, Solubility, and Miscibility

Solvents and solutes can be broadly classified into *polar* (hydrophilic) and *non-polar* (lipophilic). The polarity can be measured as the dielectric constant or the dipole moment of a compound. The polarity of a solvent determines what type of compounds it is able to dissolve and with what other solvents or liquid compounds it is miscible.

As a rule of thumb, polar solvents dissolve polar compounds best and non-polar solvents dissolve non-polar compounds best: "like dissolves like". Strongly polar compounds like inorganic salts (e.g. table salt) or sugars (e.g. sucrose) dissolve only in very polar solvents like water, while strongly non-polar compounds like oils or waxes dissolve only in very non-polar organic solvents like hexane. Similarly, water and hexane (or vinegar and vegetable oil) are not miscible with each other and will quickly separate into two layers even after being shaken well.

Protic and Aprotic Solvents

Polar solvents can be further subdivided into polar protic solvents and polar aprotic solvents. A polar protic solvent is one that contains an O-H or N-H bond.

A polar aprotic solvent is one that does not contain an O-H or N-H bond. Water (H-O-H), ethanol (CH_3-CH_2-OH), or acetic acid (CH_3-C(=O)OH) are representative polar protic solvents. A polar aprotic solvent is acetone (CH_3-C(=O)-CH_3).

Boiling Point

Another important property of solvents is boiling point. This also determines the speed of evaporation. Small amounts of low-boiling solvents like diethylether, dichloromethane, or acetone will evaporate in seconds at room temperature, while high-boiling solvents like water or dimethylsulphoxide need higher temperatures, an air flow, or the application of vacuum for fast evaporation.

Density

Most organic solvents have a lower density than water, which means they are lighter and will form a separate layer on top of water. An important exception: many halogenated solvents like dichloromethane or chloroform will sink to the bottom of a container, leaving water as the top layer.

This is important to remember when partitioning compounds between solvents and water in a separatory funnel during chemical synthesis.

Chemical Interactions

A solvent will create various weak chemical interactions with the solute to solubilise the solute. The most usual of these interactions are the relatively weak van der Waals interactions (induced dipole interactions), the stronger dipole-dipole interactions, and even the strongest interaction, hydrogen bonds (interaction between O-H or N-H hydrogens with adjacent O or N atoms).

Health and Safety

Fire: Most organic solvents are flammable or highly flammable, depending on their volatility. Exceptions are some chlorinated solvents like dichloromethane and chloroform. Mixtures of solvent vapours and air can explode.

Solvent vapours are heavier than air, they will sink to the bottom and can travel large distances nearly undiluted. Solvent vapours can also form in supposedly empty drums and cans, posing a flash fire hazard; hence empty containers of volatile solvents should be stored open and upside down.

Both diethyl ether and carbon disulphide have exceptionally low autoignition temperatures which increase greatly the fire risk associated with these solvents. The autoignition temperature of carbon disulphide is below 100°C (212°F), so as a result objects such as steam pipes, light bulbs, hotplates and recently extinguished bunsen burners are able to ignite its vapours.

Peroxide Formation

Ethers like diethyl ether and tetrahydrofuran (THF) can form highly explosive organic peroxides upon exposure to oxygen and light, THF is normally more able to form such peroxides than diethyl ether. One of the most susceptible solvents is diisopropyl ether.

The heteroatom (oxygen) stabilises the formation of a free radical which is formed by the abstraction of a hydrogen atom by another free radical. The carbon centred free radical thus formed is able to react with an oxygen molecule to form a peroxide compound. A range of tests can be used to detect the presence of a peroxide in an ether, one is to use a combination of iron sulphate and potassium thiocyanate.

The peroxide is able to oxidise the ferrous ion to a ferric ion which then form a deep red coordination complex with the thiocyanate. In extreme cases the peroxides can form crystalline solids within the vessel of the ether.

Unless the desiccant used can destroy the peroxides, they will concentrate during distillation due to their higher boiling point. When sufficient peroxides have formed, they can form a crystalline and shock sensitive solid which precipitates.

When this solid is formed at the mouth of the bottle, turning the cap may provide sufficient energy for the peroxide to detonate.

Peroxide formation is not a significant problem when solvents are used up quickly; they are more of a problem for laboratories which take years to finish a single bottle. Ethers have to be stored in the dark in closed canisters in the presence of stabilisers like butylated hydroxytoluene (BHT) or over sodium hydroxide.

Peroxides may be removed by washing with acidic ferrous sulphate, filtering through alumina, or distilling from sodium/ benzophenone. Alumina does not destroy the peroxides; it merely traps them. The advantage of using sodium/ benzophenone is that moisture and oxygen is removed as well.

Health Effects

Many solvents can lead to a sudden loss of consciousness if inhaled in large amounts. Solvents like diethylether and chloroform have been used in medicine as anesthetics, sedatives, and hypnotics for a long time.

Ethanol is a widely used and abused psychoactive drug. Diethylether, chloroform, and many other solvents (e.g. from gasoline or glues) are used recreationally in glue sniffing, often with harmful long term health effects like neurotoxicity or cancer. Methanol can cause internal damage to the eyes, including permanent blindness.

It is interesting to note that ethanol has a synergistic effect when taken in combination with many solvents. For instance a combination of toluene/benzene and ethanol causes greater nausea/vomiting than either substance alone.

Many chemists make a point of not drinking beer/wine/

other alcoholic drinks if they know that they have been exposed to an aromatic solvent.

Environmental Contamination

A major pathway to induce health effects arises from spills or leaks of solvents that reach the underlying soil. Since solvents readily migrate substantial distances, the creation of widespread soil contamination is not uncommon; there may be about 5000 sites worldwide that have major subsurface solvent contamination; this is particularly a health risk if aquifers are affected.

Chronic Health Effects

Some solvents including chloroform and benzene (an ingredient of gasoline) are carcinogenic. Many others can damage internal organs like the liver, the kidneys, or the brain.

General Precautions

- Avoiding the generation of solvent vapours by working in a fume hood, local exhaust ventilation (LEV), or a well ventilated area.
- Keeping the storage containers tightly closed.
- Never using open flames near flammable solvents, use of electrical heating instead.
- Never flushing flammable solvents down the drain to avoid explosions and fires.
- Avoiding the inhalation of solvent vapours.
- Avoiding contact of the solvent with the skin — many solvents are easily absorbed through the skin. They also tend to dry the skin and may cause sores and wounds.

Properties Table of Common Solvents

The solvents are grouped into non-polar, polar aprotic, and polar protic solvents and ordered by increasing polarity. The polarity is given as the dielectric constant. The density of non-polar solvents that are heavier than water is bolded.

Solvent	*Chemical Formula*	*Boiling point*	*Dielectric constant*	*Density*
Non-Polar Solvents				
Hexane	CH_3-CH_2-CH_2-CH_2-CH_2-CH_3	69°C	2.0	0.655 g/ml
Benzene	C_6H_6	80°C	2.3	0.879 g/ml
Toluene	C_6H_5-CH_3	111°C	2.4	0.867 g/ml
Diethyl ether	CH_3CH_2-O-CH_2-CH_3	35°C	4.3	0.713 g/ml
Chloroform	$CHCl_3$	61°C	4.8	1.498 g/ml
Ethyl acetate	CH_3-C(=O)-O-CH_2-CH_3	77°C	6.0	0.894 g/ml
Dichloromethane (DCM)	CH_2Cl_2	40°C	9.1	1.326 g/ml
Polar Aprotic Solvents				
1,4-Dioxane	-CH_2-CH_2-O-CH_2-CH_2-O-	101°C	2.3	1.033 g/ml
Tetrahydrofuran (THF)	-CH_2-CH_2-O-CH_2-CH_2-	66°C	7.5	0.886 g/ml
Acetone	CH_3-C(=O)-CH_3	56°C	21	0.786 g/ml
Acetonitrile (MeCN)	CH_3-CN	82°C	37	0.786 g/ml
Dimethylformamide (DMF)	H-C(=O)N$(CH_3)_2$	153°C	38	0.944 g/ml
Dimethyl sulphoxide (DMSO)	CH_3-S(=O)-CH_3	189°C	47	1.092 g/ml
Polar Protic Solvents				
Acetic acid	CH_3-C(=O)OH	118°C	6.2	1.049 g/ml
n-Butanol	CH_3-CH_2-CH_2-CH_2-OH	118°C	18	0.810 g/ml
Isopropanol (IPA)	CH_3-CH(-OH)-CH_3	82°C	18	0.785 g/ml
n-Propanol	CH_3-CH_2-CH_2-OH	97°C	20	0.803 g/ml
Ethanol	CH_3-CH_2-OH	79°C	24	0.789 g/ml
Methanol	CH_3-OH	65°C	33	0.791 g/ml
Formic acid	H-C(=O)OH	100°C	58	1.21 g/ml
Water	H-O-H	100°C	80	1.000 g/ml

Glycol Ethers

Glycol ethers are a group of solvents based on alkyl ethers of ethylene glycol, also sometimes called Cellosolve.

These solvents typically have higher boiling point, together with the favourable solvent properties of lower molecular weight ethers and alcohols. The original glycol ether is ethyl cellosolve.

Glycol ethers can be also derived of diethylene glycol (carbitols). Acetates of glycols are a similar kind of potent solvents.

Glycol ether solvents include:

- Ethylene glycol monomethyl ether (2-methoxyethanol, $CH_3OCH_2CH_2OH$)
- Ethylene glycol monoethyl ether (2-ethoxyethanol, $CH_3CH_2OCH_2CH_2OH$)
- Ethylene glycol monopropyl ether (2-propoxyethanol, $CH_3CH_2CH_2OCH_2CH_2OH$)
- Ethylene glycol monoisopropyl ether (2-isopropoxyethanol, $(CH_3)_2CHOCH_2CH_2OH$)
- Ethylene glycol monobutyl ether (2-butoxyethanol, $CH_3CH_2CH_2CH_2OCH_2CH_2OH$), a widely used solvent in paintings and surface coatings, cleaning products and inks
- Ethylene glycol mono phenyl ether (2-phenoxyethanol, $C_6H_5OCH_2CH_2OH$)
- Ethylene glycol monobenzyl ether (2-benzyloxyethanol, $C_6H_5CH_2OCH_2CH_2OH$)
- Diethylene glycol monoethyl ether (2-(2-ethoxyethoxy) ethanol, carbitol cellosolve,

  $CH_3CH_2OCH_2CH_2OCH_2CH_2OH$)
- Diethylene glycol mono-n-butyl ether (2-(2-butoxyethoxy) ethanol,

  $CH_3CH_2CH_2CH_2OCH_2CH_2OCH_2CH_2OH$)

Dialkyl Ethers

- Ethylene glycol dimethyl ether (dimethoxyethane, $CH_3OCH_2CH_2OCH_3$), a higher boiling alternative to diethyl ether and THF, also used as a solvent for polysaccharides, a reagent in organometallic chemistry and in some electrolytes of lithium batteries.
- Ethylene glycol diethyl ether (diethoxyethane, $CH_3CH_2OCH_2CH_2OCH_2CH_3$)
- Ethylene glycol dibutyl ether (dibutoxyethane, $CH_3CH_2CH_2CH_2OCH_2CH_2OCH_2CH_2CH_2CH_3$)

Esters

- Ethylene glycol methyl ether acetate (2-methoxyethyl acetate, $CH_3OCH_2CH_2OCOCH_3$)
- Ethylene glycol monethyl ether acetate (2-ethoxyethyl acetate, $CH_3CH_2OCH_2CH_2OCOCH_3$)
- Ethylene glycol monobutyl ether acetate (2-butoxyethyl acetate, $CH_3CH_2CH_2CH_2OCH_2CH_2OCOCH_3$)

2-Butoxyethanol

2-Butoxyethanol, which has many names, is an organic solvent with the formula $CH_3CH_2CH_2CH_2OCH_2CH_2OH$. It is a colourless liquid with a sweet, ether-like odour.

Production

In 2006, the total European production of all butyl glycol ethers amount to 181 kt/a. Approximately 50 per cent of this is 2-butoxyethanol, viz., 90 kt/a. World production is estimated to be 200 to 500 kt/a, of which 75 per cent for paints and coatings.

Main producers include:

- ICI
- Union Carbide
- BP Chemicals
- Shell Chemicals

Uses

The main use of 2-butoxyethanol is as a solvent in paints and surface coatings, followed by cleaning products and inks. Other products which contain 2-butoxyethanol include acrylic resin formulations, asphalt release agents, firefighting foam, leather protectors, oil spill dispersants and photographic strip solutions. 2-Butoxyethanol is a primary ingredient of various whiteboard cleaners, liquid soaps, cosmetics, dry cleaning solutions, lacquers, varnishes, herbicides, and latex paints. It also seems to be excellent at killing most insects and arachnids.

It is the main ingredient of many home, commercial, and industrial cleaning solutions, including the Clorox Company's trademarked Formula 409 cleaning product.

Environmental Impact

Butoxyethanol usually decomposes in the environment within a few days and has not been identified as a major environmental contaminant. It is not known to build up in any plant or animal species.

Dimethoxyethane

Dimethoxyethane, also known as glyme, monoglyme, dimethyl glycol, ethylene glycol dimethyl ether, dimethyl cellosolve, and DME, is a clear, colourless, aprotic, and liquid ether that is used as a solvent. Dimethoxyethane is highly soluble in water. Dimethoxyethane is often used as a higher boiling alternative to diethyl ether and THF. Dimethoxyethane forms chelate complexes with cations and acts as a bidentate ligand. It is therefore often used in organometallic chemistry like Grignard reactions, hydride reductions, and palladium catalysed reactions like Suzuki reactions and Stille coupling. Dimethoxyethane is also a good solvent for oligo- and polysaccharides.

Together with a high-permittivity chemical (e.g. propylene carbonate), dimethoxyethane is used as the low-viscosity component of the solvent for electrolytes of lithium batteries.

2-Ethoxyethanol

2-Ethoxyethanol, also known by the trademark Cellosolve or ethyl cellosolve, is a solvent used widely in commercial and industrial applications. It is a clear, colourless, nearly odourless liquid that is miscible with water, ethanol, diethyl ether, acetone, and ethyl acetate.

2-Ethoxyethanol can be manufactured by the reaction of ethylene oxide with ethanol.

As with other glycol ethers, 2-ethoxyethanol has the useful property of being able to dissolve chemically diverse compounds. It will dissolve oils, resins, grease, waxes, nitrocellulose, and lacquers. This is an ideal property as a multi-purpose cleaner and therefore 2-ethoxyethanol is used in products such as varnish removers and degreasing solutions.

2-(2-Ethoxyethoxy)ethanol

2-(2-Ethoxyethoxy)ethanol, also known under trade names Carbitol, Carbitol cellosolve, Transcutol, Dioxitol, Polysolve DE, and Dowanal DE, is an industrial solvent. It is a clear, colourless, hygroscopic liquid. Structurally it is an alcohol and an ether, a triethylene glycol with missing one hydroxyl group with formula $CH_3CH_2OCH_2CH_2OCH_2CH_2OH$. At direct contact it causes drying of skin by leaching fats, and is mildly irritating to the eyes. It is flammable.

2-Methoxyethanol

2-Methoxyethanol, or methyl cellosolve, is an organic compound that is used mainly as a solvent. It is a clear, colourless liquid with an ether-like odour. It is in a class of solvents known as glycol ethers which are notable for their ability to dissolve a variety of different types of chemical compounds and for their miscibility with water and other solvents.

2-Methoxyethanol is used as a solvent for many different purposes such as varnishes, dyes, and resins. It is also used as an additive jet deicing solutions.

2-Methoxyethanol is toxic to the bone marrow and testicles. Workers exposed to high levels are at risk for granulocytopenia, macrocytic anaemia, oligospermia, and azoospermia.

Phenoxyethanol

Phenoxyethanol is an organic chemical compound, a glycol ether often used in dermatological products such as skin creams. It is a colourless oily liquid. It is a bactericide (usually used in conjunction with quaternary ammonium compounds), often used in place of sodium azide in biological buffers as 2-phenoxyethanol is less toxic and non-reactive with copper and lead. It is also used as a fixative for perfumes, an insect repellent, a topical antiseptic, a solvent for cellulose acetate, some dyes, inks, and resins, in preservatives, pharmaceuticals, and in organic synthesis. It is moderately soluble in water.

Acetamide

Acetamide (or acetic acid amide or ethanamide), CH_3CONH_2, the amide of acetic acid, is a white crystalline solid in pure form. It is produced by dehydrating ammonium acetate. It is used as a plasticiser and in the synthesis of many other organic compounds.

Acetamide is not extremely combustible, but releases irritating fumes when ignited. It is toxic by inhalation (of dust), ingestion, skin and eye contact. Skin or eye contact may cause redness and pain.

The derivative N,N-dimethylacetamide (DMA), which has two methyl groups replacing the amine protons, is used as a solvent. N-methylacetamide is often used as the simplest model in studies of the peptide bond.

Recent work on the Robert C. Byrd Green Bank Telescope has resulted in the discovery of several organic (carbon-based) compounds near the centre of the Milky Way galaxy. Acetamide has been detected. This is particularly important as acetamide has an amide bond, similar to the essential bond between amino acids in proteins. This supports the theory that organic molecules that can lead to life (as we know it on Earth) can form in space.

Cancer Link

Acetamide has been found to cause cancer in laboratory animals. It is classified in Group 2B "possible human carcinogen" by the IARC.

Acetic acid

Acetic acid, also known as ethanoic acid, is an organic chemical compound with the formula CH_3COOH best recognised for giving vinegar its sour taste and pungent smell. Pure, water-free acetic acid (*glacial acetic acid*) is a colourless liquid that attracts water from the environment (hygroscopy), and freezes below 16.7°C (62°F) to a colourless crystalline solid. Acetic acid is corrosive, and its vapour causes irritation to the eyes, a dry and burning nose, sore throat and congestion to the lungs, however, it is considered a weak acid due to its limited ability to dissociate in aqueous solutions.

Acetic acid is one of the simplest carboxylic acids (the second-simplest, next to formic acid). It is an important chemical reagent and industrial chemical that is used in the production of polyethylene terephthalate mainly used in soft drink bottles; cellulose acetate, mainly for photographic film; and polyvinyl acetate for wood glue, as well as many synthetic fibres and fabrics. In households diluted acetic acid is often used in descaling agents. In the food industry acetic acid is used under the food additive code E260 as an acidity regulator.

The global demand of acetic acid is around 6.5 million tons per year (Mt/a), of which approximately 1.5 Mt/a is met by recycling; the remainder is manufactured from petrochemical feedstocks or from biological sources.

Nomenclature

The trivial name *acetic acid* is the most commonly used and officially preferred name by the IUPAC. This name derives from *acetum*, the Latin word for vinegar. The synonym *ethanoic acid* is a systematic name that is sometimes used in introductions to chemical nomenclature. Glacial acetic acid is a trivial name for water-free acetic acid. Similar to the German name *Eisessig* (literally, ice-vinegar), the name comes from the ice-like crystals that form slightly below room temperature at 16.7°C (about 62°F).

The most common and official abbreviation for acetic acid is AcOH or HOAc, where Ac stands for the acetyl group CH_3-C(=O)-. In the context of acid-base reactions the abbreviation HAc is often used where Ac instead stands for the acetate anion (CH_3COO^-), although this use is regarded by many as misleading. In either case, the Ac is not to be confused with the abbreviation for the chemical element actinium.

Acetic acid has the empirical formula CH_2O and the molecular formula $C_2H_4O_2$. The latter is often written as CH_3-COOH, CH_3COOH, or CH_3CO_2H to better reflect its structure. The ion resulting from loss of H^+ from acetic acid is the *acetate* anion. The name *acetate* can also refer to a salt containing this anion, or an ester of acetic acid.

History

Vinegar is as old as civilization itself, perhaps older. Acetic acid-producing bacteria are present throughout the world, and any culture practising the brewing of beer or wine inevitably discovered vinegar as the natural result of these alcoholic beverages being exposed to air.

The use of acetic acid in chemistry extends into antiquity. In the 3rd century BC, the Greek philosopher Theophrastos described how vinegar acted on metals to produce pigments useful in art, including *white lead* (lead carbonate) and *verdigris,* a green mixture of copper salts including copper (II) acetate. Ancient Romans boiled soured wine in lead pots to produce a highly sweet syrup called *sapa*. Sapa was rich in lead acetate, a sweet substance also called *sugar of lead* or *sugar of Saturn,* which contributed to lead poisoning among the Roman aristocracy. The 8th century Persian alchemist Jabir Ibn Hayyan (Geber) concentrated acetic acid from vinegar through distillation.

In the Renaissance, glacial acetic acid was prepared through the dry distillation of metal acetates. The 16th century German alchemist Andreas Libavius described such a procedure, and he compared the glacial acetic acid produced by this means to vinegar. The presence of water in vinegar has such a profound effect on acetic acid's properties that for centuries many chemists believed that glacial acetic acid and the acid found in vinegar were two different substances. The French chemist Pierre Adet proved them to be identical.

In 1847, the German chemist Hermann Kolbe synthesised acetic acid from inorganic materials for the first time. This reaction sequence consisted of chlorination of carbon disulphide to carbon tetrachloride, followed by pyrolysis to tetrachloroethylene and aqueous chlorination to trichloroacetic acid, and concluded with electrolytic reduction to acetic acid. By 1910 most glacial acetic acid was obtained from the "pyroligneous liquor" from distillation of wood. The acetic acid was isolated from this by treatment with milk of lime, and the resultant calcium acetate was then acidified with sulphuric acid to recover acetic acid. At this time Germany was producing 10,000 tons of glacial acetic acid, around 30 per cent of which was used for the manufacture of indigo dye.

Chemical Properties

Acidity: The hydrogen (H) atom in the carboxyl group (COOH) in carboxylic acids such as acetic acid can be given off as an H^+ ion (proton), giving them their acidic character. Acetic acid is a weak, effectively monoprotic acid in aqueous solution, with a pK_a value of 4.8. Its conjugate base is acetate (CH_3COO^-). A 1.0 M solution (about the concentration of domestic vinegar) has a pH of 2.4, indicating that merely 0.4 per cent of the acetic acid molecules are dissociated.

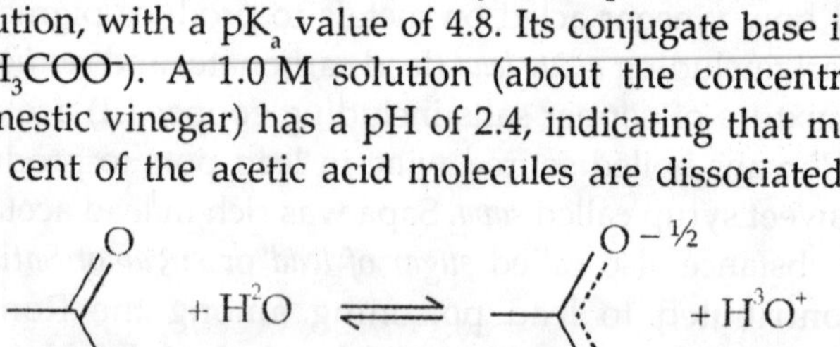

$$CH_3C(=O)OH + H^2O \rightleftharpoons CH_3C(O^{-1/2})(O^{-1/2}) + H^3O^+$$

acetic acid — acetate

Cyclic Dimer

The crystal structure of acetic acid shows that the molecules pair up into dimers connected by hydrogen bonds. The dimers can also be detected in the vapour at 120°C. They also occur in the liquid phase in dilute solutions in non-hydrogen-bonding solvents, and to some extent in pure acetic acid, but are disrupted by hydrogen-bonding solvents. The dissociation enthalpy of the dimer is estimated at 65.0-66.0 kJ/mol, and the dissociation entropy at 154-157 J mol^{-1} K^{-1}. This dimerisation behaviour is shared by other lower carboxylic acids.

Solvent

Liquid acetic acid is a hydrophilic (polar) protic solvent, similar to ethanol and water. With a moderate dielectric constant of 6.2, it can dissolve not only polar compounds such as inorganic salts and sugars, but also non-polar compounds such as oils and elements such as sulphur and iodine. It readily mixes with many other polar and non-polar solvents such as water, chloroform, and hexane. This dissolving property and miscibility of acetic acid makes it a widely used industrial chemical.

Chemical Reactions

Acetic acid is corrosive to many metals including iron, magnesium, and zinc, forming hydrogen gas and metal salts called acetates. Aluminium, when exposed to oxygen, forms a thin layer of aluminium oxide on its surface which is relatively resistant, so that aluminium tanks can be used to transport acetic acid. Metal acetates can also be prepared from acetic acid and an appropriate base, as in the popular "baking soda + vinegar" reaction. With the notable exception of chromium (II) acetate, almost all acetates are soluble in water.

$$Mg(s) + 2\ CH_3COOH(aq) \longrightarrow (CH_3COO)_2Mg(aq) + H_2(g)$$

$$NaHCO_3(s) + CH_3COOH(aq) \longrightarrow CH_3COONa(aq) + CO_2(g) + H_2O(l)$$

$$CH_3COCl \xleftarrow{SOCl_2} CH_3COOH \xrightarrow[\text{(ii) } H_2O]{\text{(i) } LiAlH_4\text{, ether}} CH_3CH_2OH$$

Acetic acid undergoes the typical chemical reactions of a carboxylic acid, such producing ethanoic acid when reacting with alkalis, producing a metal ethanoate when reacted with a metal, and producing a metal ethanoate, water and carbon dioxide when reacting with carbonates and hydrogencarbonates. Most notable of all its reactions is the formation of ethanol by reduction, and formation of derivatives such as acetyl chloride via nucleophilic acyl substitution. Other substitution derivatives include acetic anhydride; this anhydride is produced by loss of water from two molecules of acetic acid. Esters of acetic acid can likewise be formed via Fischer esterification, and amides can also be formed. When heated above 440°C, acetic acid decomposes to produce carbon dioxide and methane, or to produce ketene and water.

Detection

Acetic acid can be detected by its characteristic smell. A colour reaction for salts of acetic acid is iron chloride solution, which results in a deeply red colour that disappears after acidification. Acetates when heated with arsenic trioxide form cacodyl oxide, which can be detected by its malodorous vapours.

Biochemistry

The acetyl group, derived from acetic acid, is fundamental to the biochemistry of virtually all forms of life. When bound to co-enzyme A it is central to the metabolism of carbohydrates and fats. However, the concentration of free acetic acid in cells is kept at a low level to avoid disrupting the control of the pH of the cell contents. Unlike some longer-chain carboxylic acids (the fatty acids), acetic acid does not occur in natural triglycerides. However, the artificial triglyceride triacetin (glycerin triacetate) is a common food additive, and is found in cosmetics and topical medicines.

Acetic acid is produced and excreted by certain bacteria, notably the *Acetobacter* genus and *Clostridium acetobutylicum*. These bacteria are found universally in foodstuffs, water, and soil, and acetic acid is produced naturally as fruits and some other foods spoil. Acetic acid is also a component of the vaginal lubrication of humans and other primates, where it appears to serve as a mild anti-bacterial agent.

Production

Acetic acid is produced both synthetically and by bacterial fermentation. Today, the biological route accounts for only about 10 per cent of world production, but it remains important for vinegar production, as many of the world food purity laws stipulate that vinegar used in foods must be of biological origin. About 75 per cent of acetic acid made for use in the chemical industry is made by methanol carbonylation. Alternative methods account for the rest.

Total worldwide production of virgin acetic acid is estimated at 5 Mt/a (million tons per year), approximately half of which is produced in the United States. European production stands at approximately 1 Mt/a and is declining, and 0.7 Mt/a is produced in Japan. Another 1.5 Mt are recycled each year, bringing the total world market to 6.5 Mt/a. The two biggest producers of virgin acetic acid are Celanese and BP Chemicals. Other major producers include Millennium Chemicals, Sterling Chemicals, Samsung, Eastman, and Svensk Etanolkemi.

Methanol Carbonylation

Most virgin acetic acid is produced by methanol carbonylation. In this process, methanol and carbon monoxide react to produce acetic acid according to the chemical equation:

$$CH_3OH + CO \longrightarrow CH_3COOH$$

The process involves iodomethane as an intermediate, and occurs in three steps. A catalyst, usually a metal complex, is needed for the carbonylation.

1. $CH_3OH + HI \longrightarrow CH_3I + H_2O$
2. $CH_3I + CO \longrightarrow CH_3COI$
3. $CH_3COI + H_2O \longrightarrow CH_3COOH + HI$

By altering the process conditions, acetic anhydride may also be produced on the same plant. Because both methanol and carbon monoxide are commodity raw materials, methanol carbonylation long appeared to be an attractive method for acetic acid production. Henry Drefyus at British Celanese developed a methanol carbonylation pilot plant as early as 1925. However, a lack of practical materials that could contain the corrosive reaction mixture at the high pressures needed (200 atm or more) discouraged commercialisation of these routes for some time.

The first commercial methanol carbonylation process, which used a cobalt catalyst, was developed by German chemical company BASF in 1963. In 1968, a rhodium-based catalyst (*cis*-$[Rh(CO)_2I_2]^-$) was discovered that could operate efficiently at lower pressure with almost no by-products. The first plant using this catalyst was built by US chemical company Monsanto in 1970, and rhodium-catalysed methanol carbonylation became the dominant method of acetic acid production. In the late 1990s, the chemicals company BP Chemicals commercialised the Cativa catalyst ($[Ir(CO)_2I_2]^-$), which is promoted by ruthenium. This iridium-catalysed process is greener and more efficient and has largely supplanted the Monsanto process, often in the same production plants.

Acetaldehyde Oxidation

Prior to the commercialisation of the Monsanto process, most acetic acid was produced by oxidation of acetaldehyde. This remains the second most important manufacturing method, although it is uncompetitive with methanol carbonylation. The acetaldehyde may be produced via oxidation of butane or light naphtha, or by hydration of ethylene.

When butane or light naphtha is heated with air in the presence of various metal ions, including those of manganese, cobalt and chromium; peroxides form and then decompose to produce acetic acid according to the chemical equation:

$$2\ C_4H_{10} + 5\ O_2 \longrightarrow 4\ CH_3COOH + 2\ H_2O$$

Typically, the reaction is run at a combination of temperature and pressure designed to be as hot as possible while still keeping the butane a liquid. Typical reaction conditions are 150°C and 55 atm. Several side products may also form, including butanone, ethyl acetate, formic acid, and propionic acid. These side products are also commercially valuable, and the reaction conditions may be altered to produce more of them if this is economically useful. However, the separation of acetic acid from these by-products adds to the cost of the process.

Under similar conditions and using similar catalysts as are used for butane oxidation, acetaldehyde can be oxidised by the oxygen in air to produce acetic acid:

$$2\ CH_3CHO + O_2 \longrightarrow 2\ CH_3COOH$$

Using modern catalysts, this reaction can have an acetic acid yield greater than 95 per cent. The major side products are ethyl acetate, formic acid, and formaldehyde, all of which have lower boiling points than acetic acid and are readily separated by distillation.

Ethylene Oxidation

Oxidative Fermentation: For most of human history, acetic acid, in the form of vinegar, has been made by bacteria of the genus *Acetobacter*. Given sufficient oxygen, these bacteria can produce vinegar from a variety of alcoholic foodstuffs. Commonly used feeds include apple cider, wine, and fermented grain, malt, rice, or potato mashes. The overall chemical reaction facilitated by these bacteria is:

$$C_2H_5OH + O_2 \longrightarrow CH_3COOH + H_2O$$

A dilute alcohol solution inoculated with *Acetobacter* and kept in a warm, airy place will become vinegar over the course of a few months. Industrial vinegar-making methods accelerate this process by improving the supply of oxygen to the bacteria.

The first batches of vinegar produced by fermentation probably followed errors in the winemaking process. If must is fermented at too high a temperature, acetobacter will overwhelm the yeast naturally occurring on the grapes. As the demand for vinegar for culinary, medical, and sanitary purposes increased, vintners quickly learned to use other organic materials to produce vinegar in the hot summer months before the grapes were ripe and ready for processing into wine. This method was slow, however, and not always successful, as the vintners did not understand the process.

One of the first modern commercial processes was the "fast method" or "German method", first practised in Germany in 1823. In this process, fermentation takes place in a tower packed with wood shavings or charcoal. The alcohol-containing feed is trickled into the top of the tower, and fresh air supplied from the bottom by either natural or forced convection. The improved air supply in this process cut the time to prepare vinegar from months to weeks.

Most vinegar today is made in submerged tank culture, first described in 1949 by Otto Hromatka and Heinrich Ebner. In this method, alcohol is fermented to vinegar in a continuously stirred tank, and oxygen is supplied by bubbling air through the solution. Using this method, vinegar of 15 per cent acetic acid can be prepared in only 2–3 days.

Anaerobic Fermentation

Some species of anaerobic bacteria, including several members of the genus *Clostridium*, can convert sugars to acetic acid directly, without using ethanol as an intermediate. The overall chemical reaction conducted by these bacteria may be represented as:

$$C_6H_{12}O_6 \longrightarrow 3CH_3COOH$$

More interestingly from the point of view of an industrial chemist, many of these acetogenic bacteria can produce acetic acid from one-carbon compounds, including methanol, carbon monoxide, or a mixture of carbon dioxide and hydrogen:

$$2\ CO_2 + 4\ H_2 \longrightarrow CH_3COOH + 2\ H_2O$$

This ability of *Clostridium* to utilise sugars directly, or to produce acetic acid from less costly inputs, means that these bacteria could potentially produce acetic acid more efficiently than ethanol-oxidisers like *Acetobacter*. However, *Clostridium* bacteria are less acid-tolerant than *Acetobacter*. Even the most acid-tolerant *Clostridium* strains can produce vinegar of only a few per cent acetic acid, compared to some *Acetobacter* strains that can produce vinegar of up to 20 per cent acetic acid. At present, it remains more cost-effective to produce vinegar using *Acetobacter* than to produce it using *Clostridium* and then concentrating it. As a result, although acetogenic bacteria have been known since 1940, their industrial use remains confined to a few niche applications.

Applications

Acetic acid is a chemical reagent for the production of many chemical compounds. The largest single use of acetic acid is in the production of vinyl acetate monomer, closely followed by acetic anhydride and ester production. The volume of acetic acid used in vinegar is comparatively small.

Vinyl Acetate Monomer

The major use of acetic acid is for the production of vinyl acetate monomer (VAM). This application consumes approximately 40 per cent to 45 per cent of the world's production of acetic acid. The reaction is of ethylene and acetic acid with oxygen over a palladium catalyst.

$$2\ H_3C\text{-}COOH + 2\ C_2H_4 + O_2 \longrightarrow 2\ H_3C\text{-}CO\text{-}O\text{-}CH{=}CH_2 + 2\ H_2O$$

Vinyl acetate can be polymerised to polyvinyl acetate or to other polymers, which are applied in paints and adhesives.

Acetic Anhydride

The condensation product of two molecules of acetic acid is acetic anhydride. The worldwide production of acetic anhydride is a major application, and uses approximately 25 per cent to 30 per cent of the global production of acetic acid. Acetic anhydride may be produced directly by methanol carbonylation bypassing the acid, and Cativa plants can be adapted for anhydride production.

$$2\,CH_3COOH \longrightarrow (CH_3CO)_2O + H_2O$$

acetic acid + acetic acid → acetic anhydride

Acetic anhydride is a strong acetylation agent. As such, its major application is for cellulose acetate, a synthetic textile also used for photographic film. Acetic anhydride is also a reagent for the production of aspirin, heroin, and other compounds.

Vinegar

In the form of vinegar, acetic acid solutions (typically 5 per cent to 18 per cent acetic acid, with the percentage usually calculated by mass) are used directly as a condiment, and also in the pickling of vegetables and other foodstuffs. Table vinegar tends to be more diluted (5 per cent to 8 per cent acetic acid), while commercial food pickling generally employs more concentrated solutions. The amount of acetic acid used as vinegar on a worldwide scale is not large, but historically, this is by far the oldest and most well-known application.

Use as Solvent

Glacial acetic acid is an excellent polar protic solvent, as noted above. It is frequently used as a solvent for recrystallisation to purify organic compounds. Pure molten acetic acid is used as a solvent in the production of terephthalic acid (TPA), the raw material for polyethylene terephthalate (PET). Although currently accounting for 5 per cent-10 per cent of acetic acid use worldwide, this specific application is expected to grow significantly in the next decade, as PET production increases.

Acetic acid is often used as a solvent for reactions involving carbocations, such as Friedel-Crafts alkylation. For example, one stage in the commercial manufacture of synthetic camphor involves a Wagner-Meerwein rearrangement of camphene to isobornyl acetate; here acetic acid acts both as a solvent and as a nucleophile to trap the rearranged carbocation. Acetic acid is the solvent of choice when reducing an aryl nitro-group to an aniline using palladium-on-carbon.

Glacial acetic acid is used in analytical chemistry for the estimation of weakly alkaline substances such as organic amides. Glacial acetic acid is a much weaker base than water, so the amide behaves as a strong base in this medium. It then can be titrated using a solution in glacial acetic acid of a very strong acid, such as perchloric acid.

Other Applications

Dilute solutions of acetic acids are also used for their mild acidity. Examples in the household environment include the use in a stop bath during the development of photographic films, and in descaling agents to remove limescale from taps and kettles. The acidity is also used for treating the sting of the box jellyfish by disabling the stinging cells of the jellyfish, preventing serious injury or death if applied immediately, and for treating outer ear infections in people in preparations such as Vosol. Equivalently, acetic acid is used as a spray-on preservative for livestock silage, to discourage bacterial and fungal growth. Glacial acetic acid is also used as a wart and verruca remover.

Several organic or inorganic salts are produced from acetic acid, including:

- Sodium acetate, used in the textile industry and as a food preservative (E262).
- Copper(II) acetate, used as a pigment and a fungicide.
- Aluminium acetate and iron acetate—used as mordants for dyes.
- Palladium(II) acetate, used as a catalyst for organic coupling reactions such as the Heck reaction.

Substituted acetic acids produced include:

- Monochloroacetic acid (MCA), dichloroacetic acid

(considered a by-product), and trichloroacetic acid. MCA is used in the manufacture of indigo dye.

- Bromoacetic acid, which is esterified to produce the reagent ethyl bromoacetate.
- Trifluoroacetic acid, which is a common reagent in organic synthesis.

Amounts of acetic acid used in these other applications together (apart from TPA) account for another 5-10 per cent of acetic acid use worldwide. These applications are, however, not expected to grow as much as TPA production.

Safety

Concentrated acetic acid is corrosive and must therefore be handled with appropriate care, since it can cause skin burns, permanent eye damage, and irritation to the mucous membranes. These burns or blisters may not appear until several hours after exposure.

Latex gloves offer no protection, so specially resistant gloves, such as those made of nitrile rubber, should be worn when handling the compound. Concentrated acetic acid can be ignited with some difficulty in the laboratory. It becomes a flammable risk if the ambient temperature exceeds 39°C (102°F), and can form explosive mixtures with air above this temperature (explosive limits: 5.4-16 per cent).

The hazards of solutions of acetic acid depend on the concentration. Solutions at more than 25 per cent acetic acid are handled in a fume hood because of the pungent, corrosive vapour. Dilute acetic acid, in the form of vinegar, is harmless. However, ingestion of stronger solutions is dangerous to human and animal life. It can cause severe damage to the digestive system, and a potentially lethal change in the acidity of the blood.

Due to incompatibilities, it is recommened to keep acetic acid away from chromic acid, ethylene glycol, nitric acid, perchloric acid, permanganates, peroxides and hydroxyls.

Acetone

The chemical compound acetone (also known as propanone, dimethyl ketone, 2-propanone, propan-2-one and β-ketopropane) is the simplest representative of the ketones. Acetone is a colourless, mobile, flammable liquid with melting point of -95.4°C and boiling point of 56.53°C. It has a relative density of 0.819 (at 0°C). It is readily soluble in water, ethanol, ether, etc., and itself serves as an important solvent.

The most familiar household use of acetone is as the active ingredient in nail polish remover. Acetone is also used to make plastic, fibres, drugs, and other chemicals.

Before the invention of the cumene process acetone was produced by the dry distillation of acetates, for example calcium acetate.

In addition to being manufactured as a chemical, acetone is also found naturally in the environment, including in small amounts in the human body.

Uses

Acetone is the strongest consumer-grade solvent available. It is ideal for thinning fibreglass resin, cleaning fibreglass tools and dissolving two-part epoxies and superglue before hardening. A heavy-duty degreaser, it is great for prepping metal before painting. Also thins polyester resins, vinyl and adhesives. Easily removes residues from glass and porcelain.

An important industrial use for acetone involves its reaction with phenol for the manufacture of bisphenol A. Bisphenol A is an important component of many polymers such as polycarbonates, polyurethanes and epoxy resins. Acetone is also used extensively for the safe transporting and storing of acetylene. Vessels containing a porous material are first filled with acetone followed by acetylene, which dissolves into the acetone. One litre of acetone can dissolve around 250 litres of acetylene.

Acetone is often the primary (or only) component in nail

polish remover. Acetonitrile, another organic solvent, is sometimes used as well. Acetone is also used as a superglue remover. It can be used for thinning and cleaning fibreglass resins and epoxies. It is a strong solvent for most plastics and synthetic fibres.

Additionally, acetone is extremely effective when used as a cleaning agent when dealing with permanent markers. Also acetone can be used as an artistic agent; when rubbed on the back of any laser print or laser photocopy it produces a rough ready effect.

Acetone has been used in the manufacture of cordite. During World War I a new process of producing acetone through bacterial fermentation was developed by Chaim Weizmann, the first President of Israel, in order to help the British war effort.

Acetone can also dissolve many plastics, including those used in consumer-targeted Nalgene bottles. Acetone is also used as a drying agent, due to the readiness with which it mixes with water, and its volatility.

In the laboratory, acetone is used as a polar aprotic solvent in a variety of organic reactions.

Another industrial application is to use it as a general purpose cleaner in paint and ink manufacturing operations.

Use as an Automotive Fuel Additive

Some automotive enthusiasts add acetone at around 1 part in 500 to their fuel, following claims of dramatic improvement in fuel economy and engine life. This practice is controversial as there are counterclaims that acetone has no measurable effect or may in fact reduce engine life by adversely affecting fuel system parts.

Health Effects

Acetone is an irritant and inhalation may lead to hepatotoxic effects (causing liver damage). The fumes should be avoided. In no circumstance should it be consumed directly or indirectly.

Always use goggles when handling acetone; it can cause permanent eye damage (corneal clouding).

Small amounts of acetone are metabolically produced in the body, mainly from fat. In humans, fasting significantly increases its endogenous production. Acetone can be elevated in diabetes. Contamination of water, food (e.g. milk), or the air (acetone is volatile) can lead to chronic exposure to acetone. A number of acute poisoning cases have been described. Relatively speaking, acetone is not a very toxic compound; it can, however, damage the mucosa of the mouth and can irritate and damage skin. Accidental intake of large amounts of acetone may lead to unconsciousness and death.

The effects of long-term exposure to acetone are known mostly from animal studies. Kidney, liver, and nerve damage, increased birth defects, and lowered reproduction ability of males (only) occurred in animals exposed long-term. It is not known if these same effects would be exhibited in humans. Pregnant women should avoid contact with acetone and acetone fumes in order to avoid the possibility of birth defects, including increased brain damage.

Interestingly, acetone has been shown to have anti-convulsant effects in animal models of epilepsy, in the absence of toxicity, when administered in millimolar concentrations. It has been hypothesised that the high fat low carbohydrate ketogenic diet used clinically to control drug-resistant epilepsy in children works by elevating acetone in the brain.

Safety

Due to incompatibilities, it is recommended to keep acetone away from bromine, chlorine, nitric acid, sulphuric acid and Trichloromethane.

Acetonitrile

Acetonitrile is the chemical compound with formula CH_3CN. This colourless liquid is the simplest organic nitrile and is widely used as a solvent.

Industrial Applications

Acetonitrile is used as a solvent but also as an intermediate in the production of many chemicals ranging from pesticides to perfumes. Production trends for acetonitrile generally follow those of acrylonitrile.

The four main producers of acetonitrile in the United States are: INEOS, DuPont, J.T. Baker Chemical, and Sterling Chemicals. In 1992, 32.3 million pounds (14,700 t) of acetonitrile were produced in the US.

Laboratory uses

Acetonitrile is commonly the solvent of choice for testing an unknown chemical reaction. It is polar with a convenient liquid range. It dissolves a wide range of compounds without complications due to the fact that it contains no acidic protons. Acetone has similar properties but is more acidic and more reactive towards bases and nucleophiles.

In inorganic chemistry, acetonitrile is widely employed as a displaceable ligand where it is abbreviated MeCN. A good example is the use of $PdCl_2(MeCN)_2$ prepared by refluxing polymeric palladium chloride in acetonitrile.

It is a popular solvent in cyclic voltammetry because of its relatively high dielectric constant. MeCN is a two-carbon building block in organic synthesis. Acetonitrile is also commonly used in column chromatography and the more modern high performance liquid chromatography where it serves as a mobile phase in the separation of molecules.

Safety

Acetonitrile is toxic and flammable. It is metabolised into hydrogen cyanide and thiocyanate.

Acetophenone

Acetophenone is a crystalline ketone that is used as a solvent for cellulose ethers and esters in the manufacture of alcohol-soluble resins.

These in turn are used in perfume. Acetophenone is used to create fragrances that resemble almond, cherry, honeysuckle, jasmine, and strawberry. It is used in chewing gum. It can be found naturally in apple, cheese, apricot, banana, beef and cauliflower.

This chemical may be obtained by the dry distillation of a mixture of the calcium salts of acetic and benzoic acids. Currently acetophenone mainly comes as a by-product of the phenol-acetone synthesis in the cumene oxidation process. At one time it was used as a hypnotic under the name of hypnone.

Acidity Function

An acidity function is a measure of the acidity of a medium or solvent system, usually expressed in terms of its ability to donate protons to (or accept protons from) a solute (Bronsted acidity). The pH scale is by far the most commonly used acidity function, and is ideal for dilute aqueous solutions. Other acidity functions have been proposed for different environments, most notably the Hammett acidity function, H_0, for superacid media and its modified version H_- for superbasic media. The term acidity function is also used for measurements made on basic systems, and the term basicity function is uncommon.

Hammett-type acidity functions are defined in terms of a buffered medium:

$$H_0 = pK_a + \log \frac{c_B}{c_{BH^+}}$$

where pK_a is the dissociation constant of BH^+. They were originally measured by using nitroanilines as acid-base indicators and by measuring the concentrations of the protonated and unprotonated forms with UV-visible spectroscopy. Other spectroscopic methods, such as NMR, may also be used. The function H_- is defined similarly:

$$H_- = pK_a + \log \frac{c_{B^-}}{c_{\mathrm{BH}}}$$

Comparison of Acidity Functions with Aqueous Acidity

In dilute aqueous solution, H_0 and H_- are equivalent to pH values determined by the buffer equation. However, an H_0 value of "21 (a 25 per cent solution of SbF_5 in HSO_3F) does not imply a hydrogen ion concentration of 10^{21} mol/dm^3: such a "solution" would have a density more than a hundred times greater than a neutron star.

Rather, H_0 = "21 implies that the reactivity of the solvated hydrogen ions is 10^{21} times greater than the reactivity of the hydrated hydrogen ions in an aqueous solution of pH 0. The actual reactive species are different in the two cases, but both can be considered to be sources of H^+, i.e. Bronsted acids.

The hydrogen ion H^+ never exists on its own in a condensed phase, it is always solvated to a certain extent. The high negative value of H_0 in SbF_5/HSO_3F mixtures indicates that the solvation of the hydrogen ion is much weaker in this solvent system than in water. Other way of expressing the same phenomenon is to say that $SbF_5 \cdot FSO_3H$ is a much stronger proton donor than $[H(OH_2)_2]^+$.

Benzaldehyde

Benzaldehyde (C_6H_5CHO) is a chemical compound consisting of a benzene ring with an aldehyde substituent. It is the simplest representative of the aromatic aldehydes and one of the most industrially used members of this family of compounds. At room temperature it is a colourless liquid with a characteristic and pleasant almond-like odour: benzaldehyde is an important component of the scent of almonds, hence its typical odour.

It is the primary component of bitter almond oil extract, and can be extracted from a number of other natural sources in which it occurs, such as apricot, cherry, and laurel leaves, peach seeds and, in a glycoside combined form (amygdalin), in certain nuts and kernels. Currently benzaldehyde is primarily made from toluene by a number of different processes.

Production

Benzaldehyde can be obtained by many processes. Currently liquid phase chlorination or oxidation of toluene are among the most used processes. There is also a number of discontinued applications such as partial oxidation of benzyl alcohol, alkali treating of benzal chloride and reaction between benzene and carbon monoxide.

Reactions

On oxidation, benzaldehyde is converted into unpleasant smelling benzoic acid. Benzyl alcohol can be formed from benzaldehyde by means of hydrogenation or by treating the compound with alcoholic potassium hydroxide thus undergoing a simultaneous oxidation and reduction which result in the production of potassium benzoate and benzyl alcohol. Reaction of benzaldehyde with anhydrous sodium acetate and acetic anhydride yields cinnamic acid, while alcoholic potassium cyanide can be used to catalyse the condensation of benzaldehyde to benzoin.

Benzaldehyde can also undergo disproportionation in concentrated alkali (Cannizzaro's reaction): one molecule of the aldehyde is reduced to the corresponding alcohol and another molecule is simultaneously oxidised to the salt of a carboxylic acid. The speed of this reaction depends on the substituents present in the aromatic ring.

While it is commonly employed as a commercial food flavourant (almond flavour) or industrial solvent, benzaldehyde is used chiefly in the synthesis of other organic compounds, ranging from pharmaceuticals to plastic additives. It is also an important intermediate for the processing of perfume and flavouring compounds and in the preparation of certain aniline dyes. The synthesis of mandelic acid starts from benzaldehyde:

First hydrocyanic acid is added to benzaldehyde and the resulting mandelic acid nitrile is subsequently hydrolysed to a racemic mixture of mandelic acid. (The scheme above depicts only one of the two formed enantiomers).

Glaciologists LaChapelle and Stillman reported in 1966 that benzaldeyde and N-heptaldehyde inhibit the recrystallisation of snow and therefore the formation of depth hoar. This treatment may prevent avalanches caused by unstable depth hoar layers. However, the chemicals are not in widespread use because they damage vegetation and contaminate water supplies.

Benzene

Benzene is an organic chemical compound with the formula C_6H_6. It is sometimes abbreviated Ph-H. Benzene is a colourless and flammable liquid with a sweet smell and a relatively high melting point. It is carcinogenic and its use as additive in gasoline is now limited, but it is an important industrial solvent and precursor in the production of drugs, plastics, synthetic rubber, and dyes. Benzene is a natural constituent of crude oil, but it is usually synthesised from other compounds present in petroleum. Benzene is an aromatic hydrocarbon and the second [*n*]-annulene ([6]-annulene), a cyclic hydrocarbon with a continuous alternation of single and double bonds.

Early uses

In the 19th and early-20th centuries, benzene was used as an after-shave lotion because of its pleasant smell. Prior to the 1920s, benzene was frequently used as an industrial solvent, especially for degreasing metal. As its toxicity became obvious, benzene was supplanted by other solvents, especially toluene (methyl benzene), which has similar physical properties but is not carcinogenic.

In 1903, Ludwig Roselius popularised the use of benzene to decaffeinate coffee. This discovery led to the production of Sanka (the letters "ka" in the brand name stand for *kaffein*). This process was later discontinued.

As a petrol additive, benzene increases the octane rating and reduces knocking. Consequently, petrol often contained several per cent benzene before the 1950s, when tetraethyl lead replaced it as the most widely-used anti-knock additive.

With the global phaseout of leaded petrol, benzene has made a comeback as a gasoline additive in some nations. In the United States, concern over its negative health effects and the possibility of benzene entering the groundwater have led to stringent regulation of petrol's benzene content, with limits typically around 1 per cent. European petrol specifications now contain the same 1 per cent limit on benzene content.

Current uses of Benzene

Today benzene is mainly used as an intermediate to make other chemicals. Its most widely-produced derivatives include styrene, which is used to make polymers and plastics, phenol for resins and adhesives (via cumene), and cyclohexane, which is used in the manufacture of Nylon. Smaller amounts of benzene are used to make some types of rubbers, lubricants, dyes, detergents, drugs, explosives, napalm and pesticides.

In laboratory research, toluene is now often used as a substitute for benzene. The solvent-properties of the two are similar but toluene is less toxic and has a wider liquid range.

Benzene has been used as a basic research tool in a variety of experiments including analysis of a two-dimensional gas.

Reactions of Benzene

+ y' ⇌ :B H Y + ⇌ —y +HB⁺

- Electrophilic aromatic substitution is a general method of derivatising benzene. Benzene is sufficiently nucleophilic that it undergoes substitution by acylium ions or alkyl carbocations to give substituted derivatives.

H³C O

O

+ $\xrightarrow{AlCl_3}$ + HCl

H³C Cl

- The Friedel-Crafts acylation is a specific example of electrophilic aromatic substitution. The reaction involves the acylation of benzene (or many other aromatic rings) with an acyl chloride using a strong Lewis acid catalyst such as aluminium chloride or iron chloride which act as a halogen carrier.

$$H_3C\text{—}Cl + C_6H_6 \xrightarrow{AlCl_3} C_6H_5CH_3 + HCl$$

- Like the Friedel-Crafts acylation, the Friedel-Crafts alkylation involves the alkylation of benzene (and many other aromatic rings) using an alkyl halide in the presence of a strong Lewis acid catalyst.
- Sulphonation.
- *Nitration:* Benzene undergoes nitration with nitronioum ions (NO_2^+) as the electrophile. Thus, warming benzene at 50-55 degrees Celsius, with a combination of concentrated sulphuric and nitric acid to produce the electrophile, gives nitrobenzene.
- *Hydrogenation:* Benzene and derivatives convert to cyclohexane and derivatives when treated with hydrogen at 450°K and 10atm of pressure with a finely divided nickel catalyst.
- Benzene is an excellent ligand in the organometallic chemistry of low-valent metals. Important examples include the sandwich and half-sandwich complexes respectively $Cr(C_6H_6)_2$ and $[RuCl_2(C_6H_6)]_2$.

Health Effects

Benzene exposure has serious health effects. Breathing high levels of benzene can result in death, while low levels can cause drowsiness, dizziness, rapid heart rate, headaches, tremors, confusion, and unconsciousness. Eating or drinking foods containing high levels of benzene can cause vomiting, irritation of the stomach, dizziness, sleepiness, convulsions, rapid heart rate, and death.

The major effects of benzene are chronic (long-term) exposure through the blood. Benzene damages the bone marrow and can cause a decrease in red blood cells, leading to anaemia. It can also cause excessive bleeding and depress the immune system, increasing the chance of infection.

Some women who breathed high levels of benzene for many months had irregular menstrual periods and a decrease in the size of their ovaries. It is not known whether benzene exposure affects the developing fetus in pregnant women or fertility in men.

Animal studies have shown low birth weights, delayed bone formation, and bone marrow damage when pregnant animals breathed benzene.

The US Department of Health and Human Services (DHHS) classifies benzene as a human carcinogen. Long-term exposure to high levels of benzene in the air can cause leukaemia, a potentially fatal cancer of the blood-forming organs. In particular, Acute myeloid leukaemia and acute lymphoblastic leukaemia (AML & ALL) may be caused by benzene.

Several tests can determine exposure to benzene. There is a test for measuring benzene in the breath; this test must be done shortly after exposure.

Benzene can also be measured in the blood; however, because benzene disappears rapidly from the blood, measurements are accurate only for recent exposures.

In the body, benzene is metabolised. Certain metabolites can be measured in the urine. However, this test must be done shortly after exposure and is not a reliable indicator of benzene exposure, since the same metabolites may be present in urine from other sources.

The US Environmental Protection Agency has set the maximum permissible level of benzene in drinking water at 0.005 milligrams per litre (0.005 mg/L). The EPA requires that spills or accidental releases into the environment of 10 pounds (4.5 kg) or more of benzene be reported to the EPA.

The US Occupational Safety and Health Administration (OSHA) has set a permissible exposure limit of 1 part of benzene per million parts of air (1 ppm) in the workplace during an 8-hour workday, 40-hour workweek. The short term exposure limit for airborne benzene is 5 ppm for 15 minutes.

In recent history there have been many examples of the harmful health effects of benzene and its derivatives. Toxic Oil Syndrome caused localised immune-suppression in Madrid in 1981 from people ingesting benzene-contaminated rapeseed oil. Chronic Fatigue Syndrome has also been highly correlated with people who eat "denatured" food that use solvents to remove fat or contain benzoic acid.

Workers in various industries that make or use benzene may be at risk for being exposed to high levels of this carcinogenic chemical. Industries that involve the use of benzene include the rubber industry, oil refineries, chemical plants, shoe manufacturers, and gasoline related industries.

In 1987, OSHA estimated that about 237,000 workers in the United States were potentially exposed to benzene, and it is not known if this number has substantially changed since then.

Water and soil contamination are important pathways of concern for transmission of benzene contact. In the US alone there are approximately 100,000 different sites which have benzene soil or groundwater contamination.

In 2005, the water supply to the city of Harbin in China with a population of almost nine million people, was cut off because of a major benzene exposure. Benzene leaked into the Songhua River, which supplies drinking water to the city, after an explosion at a China National Petroleum Corporation (CNPC) factory in the city of Jilin on 13 November.

In March 2006, the official Food Standards Agency in Britain conducted a survey of 150 brands of soft drinks. It found that four contained benzene levels above World Health Organisation limits. The affected batches were removed from sale.

Benzotrichloride

Benzotrichloride, is an organic compound which is a chlorinated derivative of toluene. It is principally used as a solvent and as an intermediate in the preparation of other chemical products such as dyes.

Benzotrichloride is produced by radical chlorination of toluene in presence or absence of phosphorus trichloride, catalysed by the light, or radical initiators such as dibenzoyl peroxide.

Benzyl Alcohol

Benzyl alcohol is an organic compound with the formula $C_6H_5CH_2OH$. The benzyl group is commonly abbreviated "Bn", thus BnOH, for benzyl alcohol. Benzyl alcohol is a colourless liquid with a mild pleasant aromatic odour.

It is a natural constituent of a variety of essential oils including jasmine, hyacinth, and ylang-ylang. Benzyl alcohol is also a useful solvent due to its polarity, low toxicity, and low vapour pressure.

Benzyl alcohol is partially soluble in water (4 g/100 mL) and completely miscible in alcohols and ether. It is prepared by the hydrolysis of benzyl chloride using sodium hydroxide:

$$C_6H_5CH_2Cl + NaOH \longrightarrow C_6H_5CH_2OH + NaCl$$

It can also be prepared via a Grignard reaction by reacting phenylmagnesium bromide (C_6H_5MgBr) with formaldehyde, followed by acidification. Like most alcohols, it reacts with carboxylic acids to form esters.

Applications

Benzyl alcohol is used as a general solvent for inks, paints, lacquers, and epoxy resin coatings. It is also a precursor to a variety of esters, used in the soap, perfume, and flavour industries, as well as for pharmaceuticals. It exhibits bacteriostatic and anti-pruritic properties. It is also used as a photographic developer.

Illustrative Organic Synthetic uses

In organic synthesis, benzyl esters are popular protecting groups because they can be removed by mild hydrogenolysis.

BnOH reacts with acrylonitrile to give N-benzylacrylamide. This is an example of a Ritter reaction:

$$C_6H_5CH_2OH + NCCHCH_2 \longrightarrow C_6H_5CH_2N(H)C(O)CHCH_2$$

Nanotechnological Uses

Benzyl alcohol has been used as a dielectric solvent for the dielectrophoretic reconfiguration of nanowires.

Health and Safety

Benzyl alcohol is used as a bacteriostatic preservative in parenteral medications. Benzyl alcohol is also known for its toxic effects including respiratory failure, vasodilation, hypotension, convulsions, and paralysis. Sixteen Neo-natal deaths have been associated with the use of benzyl alcohol as a preservative in saline flush solutions. Preservative free solutions are now being used for the infant population.

Benzyl Benzoate

Benzyl benzoate is the organic compound with the formula $C_6H_5CH_2O_2CC_6H_5$. This easily prepared compound has a variety of uses.

Synthesis

This colourless liquid is formally the condensation product of benzoic acid and benzyl alcohol. It can also be generated from benzaldehyde by the Tishchenko reaction.

$$2\ C_6H_5CHO \longrightarrow C_6H_5CH_2O_2CC_6H_5$$

Uses

Benzyl benzoate has a number of uses:

- As an anti-parasitic insecticide that kills lice and the mites responsible for the skin condition scabies
- As a fixative in fragrances to improve the stability and other characteristics of the main ingredients

- As a food additive in artificial flavours
- As a plasticiser in cellulose and other polymers
- As a solvent for various chemical reactions

Health and Safety Information

Benzyl benzoate should be avoided by people with perfume allergy.

Bis (2-ethylhexyl) Adipate

Bis (2-ethylhexyl) adipate or DEHA is a plasticiser. DEHA is an ester of 2-ethylhexanol and adipic acid. Its chemical formula is $C_{22}H_{42}O_4$. Its molecular weight is 370.57 g/mol. Its CAS number is [103-23-1] and its SMILES structure is CCCCC(CC)COC(=O)CCCCC(=O)OCC(CC)CCCC.

DEHA is sometimes called "dioctyl adipate", incorrectly. Another name for it is diisooctyl adipate.

Bis(2-ethylhexyl)adipate is also used as a functional hydraulic fluid, and a component of aircraft lubricants.

Toxicity

Suspected carcinogen. Bis (2-ethylhexyl) adipate has been demonstrated to induce liver tissue tumours in female mice.

Bis (2-ethylhexyl) Phthalate

Bis (2-ethylhexyl) phthalate (also BEHP, di-2-ethyl hexyl phthalate, DEHP, or dioctyl phthalate, DOP) is a phthalate, a branched-chain 2-ethylhexanol diester of phthalic acid. At normal temperature, it is a liquid with the viscosity similar to vegetable oil. It is soluble in oil, saliva, plasma, but not in water. It possesses good plasticising efficiency, fusion rate and viscosity.

Use

Due to its low cost, DEHP is widely used as a plasticiser in manufacturing of PVC. Plastics may contain 1 to 40 per cent of DEHP. It is also used as a hydraulic fluid and as a dielectric fluid in capacitors. DEHP also finds use as a solvent in lightsticks.

Environmental Exposure

DEHP is slowly released into the air from finished plastic, posing health risks. It can be absorbed from food and water. Higher levels have been found in milk and cheese. It can also leach into a liquid that comes in contact with the plastic; it extracts faster into non-polar solvents (e.g. oils and fats in foods packed in PVC). Food and Drug Administration (FDA) therefore permits use of DEHP-containing packaging only for foods that primarily contain water.

In soil, DEHP contamination moves very slowly. It has low water solubility, therefore leaching from disposed plastics in landfills is generally slow.

The EPA limit for DEHP in drinking water is 6 ppb. OSHA limit for occupational exposure is 5 mg/m^3 of air.

Use in Medical Devices

DEHP has been used as a plasticiser in medical devices such as intravenous tubing and bags, catheters, nasogastric tubes, dialysis bags and tubing, and blood bags and transfusion tubing, and air tubes. DEHP leaches out into the fluid or air and is transported into the patient. Certain populations seem to be particularly at risk for DEHP exposure, namely newborns in intensive care nursery settings, hemophiliacs, and kidney dialysis patients.

Metabolism

The metabolism of DEHP depends on several factors, including age, and route of absorption. Oral DEHP is preferentially hydrolysed to MEHP (mono-ethylhexyl phthalate).

Toxicity

While DEHP has a low lethal toxicity, it affects certain organs at low concentration, specifically the reproductive organs (testis and ovary), lungs, kidney, and liver. Studies show that it crosses the placenta and is embryotoxic. DEHP can also be found in breast milk. DEHP does not absorb easily through skin, therefore contact with DEHP-containing products is not likely to be harmful. DEHP is a peroxisome inducer and as such can lead to liver neoplasm in rodents, however there appear to be no data to link it to cancer in humans directly. DEHP in small amounts commonly present in environment is not considered harmful.

Risk Reduction and Alternatives

The American Academy of Pediatrics has advocated not to use medical devices that can leach DEHP into patients and, instead, to resort to DEHP-free alternatives. A number of hospitals have taken a lead to reduce or eliminate DEHP containing medical devices.

However, in September 2002 the European Union's Scientific Committee on Medicinal Products and Medical Devices published an Opinion on the use of di(2-ethylhexyl) phthalate (DEHP) in medical devices saying that it can make no recommendations to limit its use, even for the most highly exposed patients.

A further review is being undertaken by the EU Scientific Committee on Emerging and Newly Identified Health Risks (SCENIHR) which will give an Opinion in January 2007.

The European Commission has banned the use of DEHP and some other phthalates in PVC toys.

Bromoform

Bromoform ($CHBr_3$) is a pale yellowish liquid with a sweet odour similar to chloroform, a halomethane or haloform. Its refractive index is 1.595 (20°C, D). Small amounts are formed naturally by plants in the ocean. It is somewhat soluble in water and readily evaporates into the air. Most of the bromoform that enters the environment is formed as by-products when chlorine is added to drinking water to kill bacteria.

Bromoform is one of the trihalomethanes closely related with fluoroform, chloroform and iodoform.

It can be prepared by electrolysis of alcoholic solution of potassium bromide.

Uses

Only small quantities of bromoform are currently produced industrially in the United States by the haloform reaction. In the past, it was used as a solvent, sedative and flame retardant, but now it is mainly used as a laboratory reagent.

Bromoform is also used in the separation of heavy minerals (ilmenite, rutile, zircon, garnet) from others. According to the laws of buoyancy, a solid will float in a liquid if its density is less than that of the liquid. Likewise, a solid will sink if its density is more than that of the liquid. If a liquid is to be used to separate minerals according to their densities, it should have a density that is inbetween that of the minerals. Due to bromoform's relatively high density, it is commonly used for the separation of minerals. In one application of the technique, the samples are separated using the described method, collected on filter papers, washed with methanol, and then dried and weighed.

Bromomethane

The chemical compound bromomethane, commonly known as methyl bromide, is an organic halogen compound with formula CH_3Br. It is a colourless, non-flammable gas with no distinctive smell. Its chemical properties are quite similar to those of chloromethane. Trade names for bromomethane include Embafume and Terabol.

Origin

Bromomethane originates from both natural and human sources. It occurs naturally in the ocean, where it is probably formed by algae and kelp. It is also produced by certain terrestrial plants, such as members of the Brassicaceae family. It is manufactured for agricultural and industrial use by reacting methanol with hydrobromic acid.

Uses

Until its production and use was curtailed by the Montreal Protocol, it was widely used as a soil sterilant, mainly for production of seed but also for some crops such as strawberries. In seed production, unlike crop production, it is of vital importance to avoid contaminating the crop with off-type seed of the same species. Therefore, selective herbicides cannot be used. While bromomethane is dangerous to use, it is considerably safer and more effective than the few other soil sterilants available. Its loss to the seed industry has resulted in changes to cultural practices, with increased reliance on mechanical rogueing and fallow seasons.

Bromomethane was also used as a general-purpose fumigant to kill a variety of pests including rats, insects, and fungi (and therefore also for killing 'bugs' and fungi according the IPPC esp. ISPM number 15, regulations when exporting wooden packaging to certain countries). It is also a precursor in the manufacture of other chemicals as a methylation agent, and has been used as a solvent to extract oil from seeds and wool.

While the Montreal Protocol has severely restricted the use of bromomethane internationally, the United States has successfully pushed for critical-use exemptions of the chemical. In 2004, the most recent year with available data, over 7 million pounds of bromomethane were applied to California fields, according to pesticide use statistics compiled by the California Department of Pesticide Regulation.

Ozone Depletion

Bromomethane is on the list of banned ozone-depleting substances of the Montreal Protocol. Because bromine is 60 times more destructive to ozone than chlorine, even small amounts of bromomethane cause considerable damage to the ozone layer. In 2005 and 2006, however, it was granted a critical use exemption under the Montreal Protocol.

Controversy

Bromomethane is used to prepare golf courses and sod for golf courses and elsewhere, particularly to control Bermuda grass. The Montreal Protocol stipulates that bromomethane use be phased out. The Bush Administration has adopted exceptions to prevent market disruptions.

Health Effects

If inhaled in high concentration for a short period, it produces headaches, dizziness, nausea, vomiting and weakness; this may be followed by mental excitement, convulsions and even acute mania. More prolonged inhalation of lower concentrations may cause bronchitis and pneumonia.

The liquid burns the skin, producing itching and reddening, then blisters several hours after contact. Both liquid and vapour severely damage the eyes.

Exposure levels leading to death vary from 1,600 to 60,000 ppm, depending on the duration of exposure. The respiratory, kidney, and neurologic effects are of the greatest concern to people. No cases of severe effects on the nervous system from long-term exposure to low levels have been noted in people, but studies in rabbits and monkeys have shown moderate to severe injury.

Sources and Sinks

Sources of CH_3Br include oceanic production, biomass burning, leaded fuel combustion, plant and marsh emissions, and fumigation of soils, durable goods, perishables, and structures. Sinks include photochemical decomposition in the atmosphere (reaction with hydroxyl radicals (OH) and photolysis at higher altitudes), loss to soils, chemical and biological degradation in the ocean, and uptake by green plants.

1,4-Butanediol

1,4-Butanediol is an alcohol derivative of the alkane butane, carrying two hydroxyl groups. It is a colourless viscous liquid.

Synthesis

Its industrial synthesis starts with acetylene, which is reacted with two molecules of formaldehyde to form 1,4-butynediol, a process known as the Reppe synthesis. This product is subsequently hydrogenated to give 1,4-butanediol.

Uses

1,4-Butanediol is used industrially as a solvent and in the manufacture of some types of plastics and fibres. In organic chemistry, 1,4-butanediol is used for the synthesis of α-butyrolactone (GBL). In the presence of phosphoric acid and high temperature, it dehydrates to the important solvent tetrahydrofuran. At about 200°C in the presence of a soluble ruthenium catalyst, the diol loses H_2 and forms butyrolactone.

It is also used as a recreational drug known by some users as "One Comma Four", "One Four Bee" or "One Four B-D-O". It exerts effects similar to γ-hydroxybutyrate (GHB), which is a metabolic product of 1, 4-butanediol.

Anecdotal reports indicate that 1, 4-butanediol produces a strong toxic feeling not present with GHB when ingested. These reports also indicate that it may cause damage to the liver as well as to other vital organs. Abuse has also resulted in addiction and death.

Pharmacokinetics

1,4-Butandiol is converted into GHB by the enzymes alcohol dehydrogenase and aldehyde dehydrogenase and differing levels of these enzymes may account for differences in effects and side effects between users. Because these enzymes are also responsible for metabolising alcohol there is a strong chance of a dangerous drug interaction. Emergency room patients who overdose on both alcohol and 1,4-butanediol often present with symptoms of ethanol intoxication initially and as the ethanol is metabolised the 1,4-butandiol is then able to better compete for the enzyme and a second period of intoxication ensues as the 1,4-butanediol is converted into GHB.

Pharmacodynamics

1,4-Butanediol seems to have two types of pharmacological actions. The major psychoactive effects of 1,4-butanediol are due to the fact that it is metabolised into GHB, however there is some evidence that 1,4-butanediol may have inherent alcohol-like pharmacological effects that are not due to this conversion.

Legality

While 1,4-butanediol is not currently scheduled federally in the United States, a number of states have classified 1,4-butanediol as a controlled substance and even where it has not yet been scheduled. Scheduling of 1,4-butanediol on a federal level is highly unlikely considering its common industrial use and many industrial applications.

1,2,4-Butanetriol

1,2,4-Butanetriol is a clear or slightly yellow, odourless, hygroscopic, flammable, viscous liquid. It is an alcohol with three hydrophilic alcoholic hydroxyl groups. It is similar to glycerol and erythritol. It is chiral, with two possible enantiomers.

1,2,4-Butanetriol is used in the manufacture of butanetriol trinitrate (BTTN), an important military propellant.

1,2,4-Butanetriol is also used as a precursor for two cholesterol-lowering drugs, as one of the monomers for manufacture of some polyesters, and as a solvent.

1,2,4-Butanetriol can be prepared synthetically by several different methods such as hydroformylation of glycidol and subsequent reduction of the product, sodium borohydride reduction of esterified malic acid, or catalytic hydrogenation of malic acid. However, of an increasing importance is the biotechnological synthesis using genetically engineered *Escherichia coli* and *Pseudomonas fragi* bacteria.

Butanol

Butanol or butyl alcohol (sometimes also called biobutanol when produced biologically), is a primary alcohol with a 4 carbon structure and the molecular formula of $C_4H_{10}O$. It is primarily used as a solvent, as an intermediate in chemical synthesis, and as a fuel. There are four isomeric structures for butanol.

Isomers

The unmodified term *butanol* usually refers to the straight chain isomer with the alcohol functional group at the terminal carbon, which is also known as *n*-butanol or 1-butanol. The straight chain isomer with the alcohol at an internal carbon is *sec*-butanol or 2-butanol. The branched isomer with the alcohol at a terminal carbon is isobutanol, and the branched isomer with the alcohol at the internal carbon is tert-butanol.

Butanol isomers, due to their different structures, have somewhat different melting and boiling points. All are moderately miscible in water, less so than ethanol, and more so than the higher (longer carbon chain) alcohols. This is because all alcohols have a hydroxyl group which makes them polar which in turn tends to promote solubility in water. At the same time the carbon chain of the alcohol resists solubility in water. Methanol, ethanol and propanol, are fully miscible in water because the hydroxyl group predominates while butanol is moderately miscible because of the balance between the two opposing solubility trends. Like many alcohols, butanol is toxic.

Uses

Butanol sees use as a solvent for a wide variety of chemical and textile processes, in organic synthesis and as a chemical intermediate. It is also used as a paint thinner and a solvent in other coating applications where it is used as a relatively slow evaporating latent solvent in lacquers and ambient-cured enamels. It finds other uses such as a component of hydraulic and brake fluids.

It is also used as a base for perfumes, but on its own has a highly alcoholic aroma. Butanol is also considered as a potential biofuel. Butanol at 85 per cent strength can be used in cars without any change to the engine (unlike ethanol) and it produces more power than ethanol and almost as much power as gasoline.

Salts of butanol are chemical intermediates for example alkali metal salts of tert-butanol are tert-butoxides.

Production

Since the 1950s, most butanol in the United States is produced commercially from fossil fuels. The most common process starts with propene, which is run through an hydroformylation reaction to form butanal, which is then reduced with hydrogen to butanol.

Butanol can also be produced by fermentation of biomass by bacteria. Prior to the 1950s, Clostridium acetobutylicum was used in industrial fermentation processes producing butanol. Research in the past few decades showed results of other microorganisms that can produce butanol through fermentation.

2-Butanol

2-Butanol, or sec-butanol, is a chemical compound with formula $C_4H_{10}O$. This secondary alcohol is a flammable, colourless liquid that is soluble in 12 parts water and completely miscible with polar organic solvent such as ethers and other alcohols.

Stereoisomers

2-Butanol is chiral and thus can be obtained as either of two stereoisomers designated as *(R)*-(–)-2-butanol and *(S)*-(+)-2-butanol. It is normally found as an equal mixture of the two stereoisomers — a racemic mixture.

Applications

2-Butanol is used as a solvent, as an intermediate in the manufacture of other chemicals (such as butanone), in industrial cleaners, and in paint removers. Many of the volatile esters of 2-butanol have pleasant aromas and are used as perfumes or in artificial flavours.

Manufacture

2-Butanol is manufactured industrially by the hydration of 2-butene:

$$C_4H_8 + H_2O \longrightarrow C_4H_{10}O$$

$$\text{(2-butene skeletal structure)} + H_2O \longrightarrow \text{(2-butanol skeletal structure, OH)}$$

tert-Butanol

tert-Butanol, or 2-methyl-2-propanol, is a tertiary alcohol. It is one of the four isomers of butanol. *tert*-Butanol is a clear liquid with a camphor-like odour. It is well soluble in water and miscible with ethanol and diethyl ether. It is unique among the isomers of butanol because it tends to be a solid at room temperature, with a melting point slightly above 25 degrees Celcius.

Applications

tert-Butanol is used as a solvent, as a denaturant for ethanol, as an ingredient in paint removers, as an octane booster for gasoline, as an oxygenate gasoline additive, and as an intermediate in the synthesis of other chemical commodities such as flavours and perfumes.

Preparation

tert-Butanol can be manufactured industrially by the catalytic hydration of isobutylene.

Chemistry

As a tertiary alcohol, *tert*-butanol is more stable to oxidation and less reactive than the other isomers of butanol. When *tert*-butanol is deprotonated with a strong base, the product is an alkoxide anion. In this case, it is *tert*-butoxide. For example, when *tert*-butanol is deprotonated with sodium hydride, the resultant is sodium *tert*-butoxide.

$$NaH + tBuOH \longrightarrow tBuO^{-}Na^{+} + H_2$$

The *tert*-butoxide species is itself useful as a strong, non-nucleophilic base in organic chemistry. It is able to abstract acidic protons from the substrate molecule readily, but its steric bulk prevents the group from participating in nucleophilic addition, such as in a Williamson ether synthesis.

Butanone

2-Butanone is a manufactured organic chemical. It is a colourless liquid with a sharp, sweet odour. It is a ketone, also known as methyl ethyl ketone (MEK).

2-Butanone is produced in large quantities. Nearly half of it is used in paints and other coatings because it will quickly evaporate. It dissolves many substances and is used as a solvent in processes involving gums, resins, cellulose acetate and cellulose nitratenitrocellulose coatings and in vinyl films. It is also used in the synthetic rubber industry.

It is used in manufacturing plastics, textiles, in the production of paraffin wax and in household products such as lacquer, varnishes, paint remover, a denaturing agent for denatured alcohol, glues and as a cleaning agent. It is used for synthesis of methyl ethyl ketone peroxide, a catalyst for some polymerisation reactions. It is highly flammable. It is not considered a large health threat.

2-Butanone occurs as a natural product. It is made by some trees and found in some fruits and vegetables in small amounts. It is also released to the air from car and truck exhausts.

Health Effects

The known health effects to people from exposure to 2-butanone are irritation of the nose, throat, skin, and eyes. There are no known cases of any humans dying from breathing 2-butanone alone. However, if 2-butanone is breathed along with other chemicals that damage health, it can increase the amount of damage that occurs.

Serious health effects in animals have been seen only at very high levels. When breathed, these effects included birth defects, loss of consciousness, and death. When swallowed, rats had nervous system effects including drooping eyelids and uncoordinated muscle movements. There was no damage to the ability to reproduce. Mice who breathed low levels for a short time showed temporary behavioural effects. Mild kidney damage was seen in animals that drank water with low levels of 2-butanone for a short time.

There are no long-term studies with animals either breathing or drinking 2-butanone.

Butyl Acetate

The chemical compound *n*-butyl acetate, also known as butyl ethanoate, is commonly used as a solvent in the production of lacquers and other products. It is also used as a synthetic fruit flavouring in foods such as candy, ice cream, cheeses, and baked goods. Butyl acetate is found in many types of fruit, where along with other chemicals it imparts characteristic flavours. Apples, especially of the Red Delicious variety, are flavoured in part by this chemical. It is a colourless flammable liquid with a sweet smell of banana. Humbrol Poly Cement also contains this chemical.

n-Butyl acetate has three isomers: isobutyl acetate, tert-butyl acetate, and sec-butyl acetate.

sec-Butyl Acetate

sec-Butyl acetate, or s-butyl acetate, is a solvent commonly used in the production of lacquers and other products. It is a colourless flammable liquid with a sweet smell.

sec-Butyl acetate has three isomers: *n*-butyl acetate, isobutyl acetate, and *tert*-butyl acetate.

tert-Butyl Acetate

tert-Butyl acetate, or *t*-butyl acetate is a colourless flammable liquid with a sweet smell. It is used as an anti-knock additive in gasoline, and as a solvent in the production of lacquers and other products.

tert-Butyl acetate has three isomers: *n*-butyl acetate, isobutyl acetate, and *sec*-butyl acetate.

Carbon dioxide

Carbon dioxide is a chemical compound, normally in a gaseous state, and is composed of one carbon and two oxygen atoms. It is often referred to by its formula CO_2. Carbon dioxide is present in the Earth's atmosphere at a concentration of approximately 000383 by volume (383 ppm) and is an important greenhouse gas due to its ability to absorb many infrared wavelengths of sunlight, and due to the length of time it stays in the atmosphere. It is also a major component of the carbon cycle. In its solid state, carbon dioxide is called dry ice. CO_2 has no liquid state at normal atmospheric pressure.

Chemical and Physical Properties

Carbon Dioxide is normally a colourless, odourless and neutral gas. When inhaled at concentrations higher than the usual atmospheric level of 300-600 ppm (Parts Per Million; by volume) it can produce a sour taste in the mouth and a stinging sensation in the nose and throat. These effects result from the gas dissolving in the mucous membranes and saliva, forming a weak solution of carbonic acid. One may notice this sensation if one attempts to stifle a burp after drinking a carbonated beverage. Amounts above 5000 ppm are considered unhealthy, and those above about 50000 ppm are considered dangerous to animal life.

Its density at standard temperature and pressure is around 1.98 kg/m^3, about 1.53 times that of air. The carbon dioxide molecule (O=C=O) contains two double bonds and has a linear shape. It has no electrical dipole. As it is fully oxidised, it is not very reactive and is non-flammable and therefore is considered neutral.

Under normal atmospheric pressure (1 atm) at -78.5°C, carbon dioxide changes directly from a solid phase to a gaseous phase through sublimation or gaseous to solid through deposition. The solid form is typically called "dry ice". Liquid carbon dioxide forms only at pressures above 4.0-5.1 atm, depending on temperature. Specifically, the triple point is 416.7 kPa at -56.6°C The critical point is 7821 kPa at 31.1°C.

History of Human Understanding

Carbon dioxide was one of the first gases to be described as a substance distinct from air. In the seventeenth century, the Flemish chemist Jan Baptist van Helmont observed that when he burned charcoal in a closed vessel, the mass of the resulting ash was much less than that of the original charcoal.

His interpretation was that the rest of the charcoal had been transmuted into an invisible substance he termed a "gas" or "wild spirit" (*spiritus sylvestre*).

The properties of carbon dioxide were studied more thoroughly in the 1750s by the Scottish physician Joseph Black. He found that limestone (calcium carbonate) could be heated or treated with acids to yield a gas he termed "fixed air." He observed that the fixed air was denser than air and did not support either flame or animal life. He also found that it would, when bubbled through an aqueous solution of lime (calcium hydroxide), precipitate calcium carbonate, and used this phenomenon to illustrate that carbon dioxide is produced by animal respiration and microbial fermentation.

In 1772, English chemist Joseph Priestley published a paper entitled Impregnating Water with Fixed Air in which he described a process of dripping sulphuric acid (or *oil of vitriol* as Priestley knew it) onto chalk in order to produce carbon dioxide and forcing the gas to dissolve by agitating a bowl of water in contact with the gas.

Carbon dioxide was first liquefied (at elevated pressures) in 1823 by Humphry Davy and Michael Faraday. The earliest description of solid carbon dioxide was given by Charles Thilorier, who in 1834 opened a pressurised container of liquid carbon dioxide, only to find that the cooling produced by the rapid evaporation of the liquid yielded a "snow" of solid CO_2.

Isolation

Carbon dioxide may be obtained from air distillation, however this yields only very small quantities of CO_2. A large variety of chemical reactions yield carbon dioxide, such as the reaction between most acids and most metal carbonates. As an example, the reaction between sulphuric acid and calcium carbonate (limestone or chalk) is depicted below:

$$H_2SO_4 + CaCO_3 \longrightarrow CaSO_4 + H_2CO_3$$

The H_2CO_3 then decomposes to water and CO_2. Such reactions are accompanied by foaming and/or bubbling. In industry such reactions are widespread because they can be used to neutralise waste acid streams.

The production of quicklime (CaO) a chemical that has widespread use, from limestone by heating at about 850°C also produces CO_2:

$$CaCO_3 \longrightarrow CaO + CO_2$$

The combustion of all carbon containing fuels, such as methane (natural gas), petroleum distillates (gasoline, diesel, kerosene, propane), but also of coal and wood, will yield carbon dioxide, and, in most cases, water. As an example the chemical reaction between methane and oxygen is given below.

$$CH_4 + 2\ O_2 \longrightarrow CO_2 + 2\ H_2O$$

Iron is reduced from its oxides with coke in blast furnace, producing pig iron and carbon dioxide:

$$2\ Fe_2O_3 + 3\ C \longrightarrow 4\ Fe + 3\ CO_2$$

Carbon dioxide can be used in chemistry to create a carboxylic acid from a Grignard reagent.

$$R\text{-}MgX + CO_2 \longrightarrow R\text{-}COOH$$

Yeast produces carbon dioxide and ethanol, also known as alcohol, in the production of wines, beers and other spirits:

$$\text{Glucose} \longrightarrow 2\ CO_2 + 2\ C_2H_5OH$$

All aerobic organisms produce CO_2 when they oxidise carbohydrates, fatty acids and proteins in the mitochondria of cells; it is the prime energy source and the main metabolic pathway in heterotroph organisms such as animals, and also a secondary energy source in phototroph organisms such as plants when not enough light is available for photosynthesis. The large amount of reactions involved are exceedingly complex and not described easily. Refer to (respiration, anaerobic respiration and photosynthesis).

Photoautotrophs (i.e. plants, cyanobacteria) utilise another modus operandi: They absorb the CO_2 from the air, and together with water, react it to form carbohydrates:

$$nCO_2 + nH_2O \longrightarrow (CH_2O)_n + nO_2$$

Carbon dioxide is soluble in water, in which it spontaneously interconverts between CO_2 and H_2CO_3 (carbonic acid). The relative concentrations of CO_2, H_2CO_3, and the deprotonated forms HCO_3^- (bicarbonate) and CO_3^{2-}(carbonate) depend on the pH. In neutral or slightly alkaline water (pH > 6.5), the bicarbonate form predominates (>50 per cent) becoming the most prevalent (>95 per cent) at the pH of seawater, while in very alkaline water (pH > 10.4) the predominant (>50 per cent) form is carbonate. The bicarbonate and carbonate forms are very soluble, such that air-equilibrated ocean water (mildly alkaline with typical pH = 8.2 – 8.5) contains about 120 mg of bicarbonate per litre.

Industrial Production

Carbon dioxide is manufactured mainly from six processes:

1. As a by-product in ammonia and hydrogen plants, where methane is converted to CO_2;
2. From combustion of carbonaceous fuels;
3. As a by-product of fermentation;
4. From thermal decomposition of $CaCO_3$;
5. As a by-product of sodium phosphate manufacture;
6. Directly from natural carbon dioxide gas wells.

Solid Carbon dioxide — "dry ice"

Solid carbon dioxide has the generic trademark "dry ice" and is a versatile cooling agent. Unlike when water passes its solid point and becomes molten, dry ice sublimes, changing directly to a gas. Its sublimation/deposition point is -78.5°C (-109.3°F). The low temperature and direct sublimation to a gas makes dry ice a very effective coolant, since it's colder than ice and leaves no moisture as it changes state. Dry ice is also inexpensive; it costs about US$2 per kilogram.

Dry ice was first observed in 1825 by the French chemist Charles Thilorier. Upon opening the lid of a large cylinder containing liquid carbon dioxide he noted much of the carbon dioxide rapidly evaporated leaving solid dry ice in the container. Throughout the next 60 years, dry ice was observed and tested by many scientists.

Production

Dry ice is readily manufactured:

1. Carbon dioxide is pressurised and refrigerated until it changes into its liquid form.
2. The pressure is reduced. When this occurs some liquid carbon dioxide vaporises, and this causes a rapid lowering of temperature of the remaining liquid carbon dioxide. The extreme cold makes the liquid solidify into a snow-like consistency.
3. The snow-like solid carbon dioxide is compressed into either small pellets or larger blocks of dry ice.

Dry ice is typically produced in two standard sizes: solid blocks and cylindrical pellets. A standard block is most common and will normally be about 30 kg. These are largely used in the shipping industry because they sublime slowly due to a relatively small surface area. The pellets are around 1 cm in diameter and can be bagged easily. This form of dry ice is more suited to small scale use, for example at grocery stores and laboratories.

Safety

Dry ice can be a dangerous substance, if used improperly. It must be handled using protective insulated gloves, because direct contact can cause frostbite. It must not be stored in a sealed container, since its sublimation produces large volumes of gaseous carbon dioxide at high pressure — a sealed container containing dry ice can fail explosively, which could cause shrapnel injuries and hearing loss. Dry ice should never be stored in a functioning freezer or refrigerator, because it is cold enough to freeze the thermostat. Also, due to its temperature, thermal contraction from dry ice can cause brittle materials such as glass or plastic to crack.

Applications

Dry ice has many applications:

- Transporting items that need to remain cold or frozen, such as food, without needing any mechanical cooling source.
- Blast cleaning.
- Freezing warts to make removal easier.
- Keeping broken or powerless refrigerators and freezers cold.
- Loosening floor tiles by shrinking and cracking them.
- Carbonating water and other liquids such as beer.
- Repelling mosquitoes and other insects.
- Creating low-sinking dense clouds of fog for dramatic effects by putting it in water and therefore accelerating sublimation.
- Freezing water in pipes with no valves or being repaired to stop leaking.
- Making ice cream.
- Minor dent repairs—dry ice can help to remove dents, by forcing a car's sheet metal to contract.

Dry Ice Blast Cleaning

One of the largest alternative uses of dry ice around the world is dry ice blast cleaning. Dry ice pellets are shot out of a jet nozzle with compressed air. This can remove residues from industrial equipment, for example ink, glue, oil, paint, mould and rubber, replacing sandblasting, steam blasting, water blasting or other (potentially environmentally damaging) solvent blasting.

Dry ice blasting involves three factors:

1. Kinetic energy

2. Thermal shock
3. Thermal kinetic energy

The kinetic energy of the dry ice pellets is transferred when it hits the surface, directly dislodging residues, as in other blasting methods. The thermal shock effect occurs when the cold dry ice hits a much warmer surface and rapid sublimation occurs. The thermal kinetic effect is the result of the rapid sublimation of the dry ice hitting the surface. These factors combine cause small "micro-explosions" of gaseous carbon dioxide where each pellet of dry ice impacts, dislodging the residue.

Solid Amorphous Carbon dioxide

An alternative form of solid carbon dioxide, an amorphous glass-like form, is possible, although not at atmospheric pressure. This form of glass, called *carbonia*, was produced by supercooling heated CO_2 at extreme pressure (40-48 GPa or about 400,000 atmospheres) in a diamond anvil. This discovery confirmed the theory that carbon dioxide could exist in a glass state similar to other members of its elemental family, like silicon (silica glass) and germanium. Unlike silica and germanium oxide glasses, however, carbonia glass is not stable at normal pressures and reverts back to gas when pressure is released.

Carbon dioxide in the Earth's Atmosphere

Carbon dioxide is present at a very small 383 ppm (.000383) of the volume of the earth's atmosphere, but it is a very powerful greenhouse gas and so has a large effect upon climate. It is also essential to photosynthesis in plants and other photoautotrophs.

Despite the low concentration, CO_2 is a very important component of the Earth's atmosphere because it absorbs infrared radiation at wavelengths of 4.26 μm (asymmetric stretching vibrational mode) and 14.99 μm (bending vibrational mode) and enhances the greenhouse effect to a great degree.

Although water vapour accounts for up to 90 per cent of the greenhouse effect, there is no real way to control the amount of water vapour in the Earth's climate system and it is short-lived in the atmosphere. In addition, water vapour is almost never considered a forcing, but rather almost always a feedback.

On the other hand, carbon dioxide is a very powerful forcing, and it also lasts far longer in the Earth's atmosphere. With a radiative forcing of about 1.5 W/m^2, it is relativly twice as powerful as the next majorly forcing greenhouse gas, methane, and relatively ten times as powerful as the third, nitrous oxide. Carbon dioxide contributes up to 12 per cent to the greenhouse effect.

The 20 year smoothed Law Dome DE02 and DE02-2 ice cores show the levels of CO_2 to have been 284.3 ppm in 1832. As of January 2007, the measured atmospheric CO_2 concentration at the Mauna Loa observatory was about 383 ppm.

Biological Role

Carbon dioxide is an end product in organisms that obtain energy from breaking down sugars, fats and amino acids with oxygen as part of their metabolism, in a process known as cellular respiration. This includes all plants, animals, many fungi and some bacteria. In higher animals, the carbon dioxide travels in the blood from the body's tissues to the lungs where it is exhaled. In plants using photosynthesis, carbon dioxide is absorbed from the atmosphere.

Role in Photosynthesis

Plants remove carbon dioxide from the atmosphere by photosynthesis, also called carbon assimilation, which uses light energy to produce organic plant materials by combining carbon dioxide and water. Free oxygen is released as gas from the decomposition of water molecules, while the hydrogen is split into its protons and electrons and used to generate chemical energy via photophosphorylation. This energy is required for the fixation of carbon dioxide in the Calvin cycle to form sugars. These sugars can then be used for growth within the plant through respiration. Carbon dioxide gas must be introduced into greenhouses to maintain plant growth, as even in vented greenhouses the concentration of carbon dioxide can fall during daylight hours to as low as 200 ppm, at which level photosynthesis is significantly reduced.

Venting can help offset the drop in carbon dioxide, but will never raise it back to ambient levels of 340ppm. Carbon dioxide supplementation is the only known method to overcome this deficiency. Direct introduction of pure carbon dioxide is ideal, but rarely done because of cost constraints.

Most greenhouses burn methane or propane to supply the additional CO_2, but care must be taken to have a clean burning system as increased levels of nitrous oxide (NO_2) result in reduced plant growth. Sensors for sulphur dioxide (SO_2) and NO_2 are expensive and difficult to maintain, accordingly most systems come with a carbon monoxide (CO) sensor under the assumption that high levels of carbon monoxide mean that significant amounts of NO_2 are being produced. Plants can potentially grow up to 50 per cent faster in concentrations of 1000ppm CO_2 when compared with ambient conditions.

Plants also emit CO_2 during respiration, so it is only during growth stages that plants are net absorbers. For example a growing forest will absorb many tons of CO_2 each year, however a mature forest will produce as much CO_2 from respiration and decomposition of dead specimens (e.g. fallen branches) as used in biosynthesis in growing plants. Regardless of this, mature forests are still valuable carbon sinks, helping maintain balance in the Earth's atmosphere.

Animal Toxicity

Carbon dioxide content in fresh air varies and is between 0.03 per cent (300 ppm) and 0.06 per cent (600 ppm), depending on location. Exhaled breath is approximately 4.5 per cent carbon dioxide. When inhaled in high concentrations (greater than 5 per cent by volume, or 50000 ppm), it is immediately dangerous to the life and health of humans and other animals.

The current threshold limit value (TLV) or maximum level

that is considered safe for healthy adults for an 8-hour work day is 0.5 per cent (5000 ppm). The maximum safe level for infants, children, the elderly and individuals with cardio-pulmonary health issues would be significantly less.

These figures are valid for carbon dioxide supplied in "pure" form. In indoor spaces occupied by humans the carbon dioxide concentration will also reach a level higher than in pure outdoor air. Concentrations higher than 1000 ppm will cause discomfort in more than 20 per cent of occupants, and the discomfort will increase with increasing CO_2 concentration. The discomfort will be caused by various gases coming from human respiration and perspiration, and not by CO_2 itself. At 2000 ppm the majority of occupants will feel a significant degree of discomfort, and many will develop nausea and headache. The CO_2 concentration between 300 and 2500 ppm is used as an indicator of indoor air quality in spaces polluted by human occupation.

Acute carbon dioxide toxicity is sometimes known as Choke damp, an old mining industry term, and was the cause of death at Lake Nyos in Cameroon, where an upwelling of CO_2-laden lake water in 1986 covered a wide area in a blanket of the gas, killing nearly 2000. The lowering of carbon dioxide in the atmosphere is largely due to absorption by plants, which convert it to sugars through photosynthesis. Phytoplankton photosynthesis absorbs dissolved CO_2 in the upper ocean and thereby promotes the absorption of CO_2 from the atmosphere.

Carbon dioxide ppm levels (CDPL) are a surrogate for measuring indoor pollutants that may cause occupants to grow drowsy, get headaches, or function at lower activity levels. To eliminate most Indoor Air Quality complaints, total indoor CDPL must be reduced to below 600. NIOSH considers that indoor air concentrations that exceed 1000 are a marker suggesting inadequate ventilation. ASHRAE recommends they not exceed 1000 inside a space. OSHA limits concentrations in the workplace to 5000 for prolonged periods. The US National Institute for Occupational Safety and Health limits brief exposures (up to ten minutes) to 3000 and considers CDPL exceeding 4000 as "immediately dangerous to life and health." People who breathe 5000 for more than half an hour show signs of acute hypercapnia, while breathing 7000 – 10000 can produce unconsciousness in only a few minutes. Accordingly, carbon dioxide, either as a gas or as dry ice, should be handled only in well-ventilated areas.

Human Physiology

CO_2 is carried in blood in three different ways. (The exact percentages vary depending whether it is arterial or venous blood).

- Most of it (about 80 per cent – 90 per cent) is converted to bicarbonate ions HCO_3^- by the enzyme carbonic anhydrase in the red blood cells.
- 5 per cent – 10 per cent is dissolved in the plasma
- 5 per cent – 10 per cent is bound to haemoglobin as carbamino compounds.

The CO_2 bound to haemoglobin does not bind to the same site as oxygen; rather it combines with the N-terminal groups on the four globin chains. However, because of allosteric effects on the haemoglobin molecule, the binding of CO_2 does decrease the amount of oxygen that is bound for a given partial pressure of oxygen.

Haemoglobin, the main oxygen-carrying molecule in red blood cells, can carry both oxygen and carbon dioxide, although in quite different ways. The decreased binding to carbon dioxide in the blood due to increased oxygen levels is known as the Haldane Effect, and is important in the transport of carbon dioxide from the tissues to the lungs. Conversely, a rise in the partial pressure of CO_2 or a lower pH will cause offloading of oxygen from haemoglobin. This is known as the Bohr Effect.

Carbon dioxide may be one of the mediators of local autoregulation of blood supply. If it is high, the capillaries expand to allow a greater blood flow to that tissue.

Bicarbonate ions are crucial for regulating blood pH. As

breathing rate influences the level of CO_2 in blood, too slow or shallow breathing causes respiratory acidosis, while too rapid breathing, hyperventilation, leads to respiratory alkalosis.

Although it is oxygen that the body requires for metabolism, it is not low oxygen levels that stimulate breathing, but is instead higher carbon dioxide levels. As a result, breathing low-pressure air or a gas mixture with no oxygen at all (e.g., pure nitrogen) leads to loss of consciousness without subjective breathing problems. This is especially perilous for high-altitude fighter pilots, and is also the reason why the instructions in commercial airplanes for case of loss of cabin pressure stress that one should apply the oxygen mask to oneself before helping others — otherwise one risks going unconscious without being aware of the imminent peril.

According to a study by the USDA, an average person's respiration generates approximately 450 litres (roughly 900 grams) of carbon dioxide per day.

Carbon dioxide Sequestering

As part of the current scientific opinion that excess amounts of carbon dioxide produced by humans in the atmosphere lead to global warming, various methods of limiting or removing the amount of carbon dioxide in the atmosphere have been suggested. Current debate on the subject mostly involves economic or political matters at a policy level.

Methods of carbon dioxide extraction/separation include:

1. Aqueous solutions
    - Amine extraction
    - High pH solutions
        - For example, Carbon dioxide reacts with dissolved CaO, to form Calcite ($CaCO_3$).
2. Adsorption
    - Molecular Sieves
    - Activated Carbon
    - Metal-organic frameworks (MOF's)

3. Solid reactants
   - Serpentine, Olivine, Quicklime
4. Membrane gas separation
5. Regenerative Carbon Dioxide Removal System (RCRS)
   - The RCRS on the space shuttle Orbiter uses a two-bed system that provides continuous removal of CO_2 without expendable products. Regenerable systems allow a shuttle mission a longer stay in space without having to replenish its sorbent canisters. Older lithium hydroxide (LiOH)-based systems, which are non-regenerable, are being replaced by regenerable metal-oxide-based systems. A metal-oxide-based system primarily consists of a metal oxide sorbent canister and a regenerator assembly. This system works by removing carbon dioxide using a sorbent material and then regenerating the sorbent material. The metal-oxide sorbent is regenerated by pumping air heated to around 200°C at 7.5 standard cubic feet per minute through its canister for 10 hours.
6. Algae Bioreactor Technology
   - Originally developed at MIT using power plant flue gas to support biodiesel feed stock, they use algae to process out the CO_2. Commercial studies have been performed on over 2000 MW of power plants in the United States since 2001. As of March 2007, this is the only commercially installed technology for CO_2 mitigation on active power plants. The largest test site for an Algae bioreactor system is connected directly to smokestack of Arizona Public Service Redhawk 1,040 megawatt power plant, producing renewable biofuels as a process by product. At commercial scale, this organic process holds the potential to "scrub" CO_2 without the considerable solid and fluid waste issues associated with other technologies

7. Underground geological storage.
8. Deep Ocean storage. At sufficiently high pressure, around 500 m depth, carbon dioxide forms a solid hydrate with water.
9. *Terra Preta:* Charcoal enhanced soils.
    - Amazon soils that are valued today for their rich agricultural abilities are found to contain charcoal that was put into the soils by Amazonians thousands of years ago. Plant and organic material converted to charcoal can be used to enhance soils and keep CO_2 out of the atmosphere for thousands of years. Oak Ridge National Laboratory has found a way to further enhance charcoal's agricultural benefits and capture more CO_2 by combining ammonia and fossil fuel exhaust to form ammonium bicarbonate in the charcoal lattices. The work by Oak Ridge National Laboratory is currently being commercialised by a corporation called EPRIDA, Inc.

Carbon Disulphide

Carbon disulphide is a colourless liquid with the formula CS_2. It has a pleasant odour that is like that of chloroform, although it is usually impure yellowish with an unpleasant odour like that of rotting radishes due to traces of other sulphurous species, such as carbonyl sulphide (COS).

Occurrence and Manufacture

Small amounts of carbon disulphide are found in gases released by volcanic eruptions or marshes. CS_2 once was manufactured by combining carbon (or coke) and sulphur at high temperatures. A lower temperature reaction, requiring only 600°C involves natural gas in the presence of kieselgel or alumina catalysts:

$$CH_4 + 1/2\ S_8 \longrightarrow CS_2 + 2\ H_2S.$$

Chemical Properties

CS_2 is structurally analogous to CO_2, but more reactive towards nucleophiles and bases and more easily reduced. This difference in reactivity can be attributed to the weaker π donorability of the sulphido centres.

Addition of Nucleophiles

Nucleophiles such as amines afford dithiocarbamates:

$$2R_2NH + CS_2 \longrightarrow [R_2NH_2^+][R_2NCS_2^-].$$

Xanthates form similarly from alkoxides:

$$RONa + CS_2 \longrightarrow [Na^+][ROCS_2^-].$$

Sodium sulphide affords trithiocarbonate:

$$Na_2S + CS_2 \longrightarrow [Na^+]_2[CS_3^{2-}].$$

Reduction

Sodium reduces CS_2 to give the heterocycle "dmit^{2-}":

$$3\ CS_2 + 4\ Na \longrightarrow Na_2C_3S_5 + Na_2S.$$

Direct electrochemical reduction affords the tetrathiooxalate anion:

$$2\ CS_2 + 2e \longrightarrow C_2S_4^{2-}.$$

Other

Chlorination of CS_2 is the route to thiophosgene.

CS_2 is a ligand for many metal complexes, forming pi complexes.

It is used to manufacture regenerated cellulose (the main ingredient of viscose rayon and cellophane), carbon tetrachloride and organic sulphur compounds including those mentioned above. It is used as flotation agents in mineral processing, and Metham sodium soil fumigant.

Health Effects

At very high levels, carbon disulphide may be life-threatening because it affects the nervous system. People, who breathed carbon disulphide near an accident involving a derailed railroad car showed changes in breathing and some chest pains, although these effects are probably due to the sulphur oxides formed in the ensuing fire.

Some workers, who breathed high levels during working hours for at least 6 months had headaches, tiredness, and trouble sleeping, in some cases even serious central and peripheral nervous system disease such as toxic encephalitis and peripheral mixed neuropathy.

Most of these data come from the viscose rayon Industry, where small amounts of H_2S may also have been present. Among workers, who breathed lower levels, some developed very slight changes in their nerves. An increased risk of cardiovascular death has also been established in several countries.

Studies in animals indicate that carbon disulphide can affect the normal functions of the brain, liver, and heart. After pregnant rats breathed carbon disulphide in the air, some of the newborn rats died or had birth defects. Liquid carbon disulphide has caused skin burns.

Carbon Tetrachloride

Carbon tetrachloride, also known by many other names is the chemical compound with the formula CCl_4. It is a reagent in synthetic chemistry and was formerly widely used in fire extinguishers and as a precursor to refrigerants. It is a colourless liquid with a "sweet" smell that can be detected at low levels.

Both carbon tetrachloride and tetrachloromethane are acceptable names under IUPAC nomenclature, depending on whether it is seen as an inorganic or an organic compound. Colloquially, it is called "carbon tet".

Production

Carbon tetrachloride was originally synthesised in 1839 by reaction of chloroform with chlorine, but the most common synthesis is now chlorination of carbon disulphide at 105 to 130°C:

$$CS_2 + 3Cl_2 \longrightarrow CCl_4 + S_2Cl_2$$

CCl_4 can also be prepared from methane via the reaction:

$$CH_4 + 4\ Cl_2 \longrightarrow CCl_4 + 4HCl$$

It is also a by-product in the synthesis of dichloromethane and chloroform.

Properties

In the carbon tetrachloride molecule, four chlorine atoms are positioned symmetrically as corners in a tetrahedral configuration joined to a carbon atom, in the centre, by single covalent bonds. Because of this symmetrical geometry, the molecule has no net dipole moment, that is CCl_4 is non-polar. As a solvent, it is well suited to dissolving other non-polar compounds, fats and oils. It is somewhat volatile, giving off vapours having a smell characteristic of other chlorinated solvents, somewhat similar to the tetrachloroethylene smell reminiscent of dry cleaners' shops.

Solid tetrachloromethane has 2 allotropes—crystaline II below - 47.5°C (225.6°K) and crystaline I above -47.5°C.

Uses

In the early 20th century, carbon tetrachloride was widely used as a dry cleaning solvent, as a refrigerant, and in fire extinguishers. However, once it became apparent that carbon tetrachloride exposure had severe adverse health effects, safer alternatives such as tetrachloroethylene were found for these applications, and its use in these roles declined from about 1940 onward. Carbon tetrachloride persisted as a pesticide to kill insects in stored grain, but in 1970, it was banned in consumer products in the United States.

Prior to the Montreal Protocol, large quantities of carbon tetrachloride were used to produce the freon refrigerants R-11 (trichlorofluoromethane) and R-12 (dichlorodifluoro-methane).

However, these refrigerants are now believed to play a role in ozone depletion and have been phased out. Carbon tetrachloride is still used to manufacture less destructive refrigerants.

Carbon tetrachloride has also been used in the detection of neutrinos.

Reactivity

Carbon tetrachloride has practically no flammability at lower temperatures. Under high temperatures in air, it forms poisonous phosgene.

Because it has no C-H bonds, carbon tetrachloride does not easily undergo free-radical reactions. Hence it is a useful solvent for halogenations either by the elemental halogen, or by a halogenation reagent such as *N*-bromosuccinimide.

In organic chemistry, carbon tetrachloride serves as a source of chlorine in the Appel reaction.

As a Solvent

It is used as a solvent in synthetic chemistry research, but because of its adverse health effects, it is no longer commonly used, and chemists generally try to substitute it with other solvents. It is sometimes useful as a solvent for infrared spectroscopy because there are no significant absorption bands > 1600 cm^{-1}. Because carbon tetrachloride does not have any hydrogen atoms, it was historically used in proton NMR spectroscopy. However, carbon tetrachloride is toxic, and its dissolving power is low. Its use has been largely superseded by deuterated solvents, which offer superior solvating properties and allow for deuterium lock by the spectrometer.

Safety

Exposure to high concentrations of carbon tetrachloride (including vapour) can affect the central nervous system, degenerate the liver and kidneys and may result (after prolonged exposure) in coma and even death. Chronic exposure to carbon tetrachloride can cause liver and kidney damage and could result in cancer. More information can be found in Material safety data sheets. Carbon tetrachloride is also both ozone-depleting and a greenhouse gas. However, since 1992 its atmospheric concentrations have been in decline for the reasons described above.

Chlorobenzene

Chlorobenzene is an aromatic organic compound with the chemical formula C_6H_5Cl. It is a colourless, flammable liquid first made in 1851 by reacting phenol and phosphorus pentachloride.

Uses

Chlorobenzene has been used in the manufacture of certain pesticides, most notably DDT by reaction with chloral (trichloroacetaldehyde). It once found use in the production of phenol. Today the major use of chlorobenzene is as an intermediate in the production of nitrochlorobenzenes and diphenyl oxide, which are important in the production of commodities such as herbicides, dyestuffs, and rubber. Chlorobenzene is also used as a high-boiling solvent in organic synthesis as well as many industrial applications.

Synthesis

Chlorobenzene is prepared by chlorination of benzene, usually in the presence of a catalytic amount of Lewis acid such as ferric chloride:

$$C_6H_6 + Cl_2 \longrightarrow C_6H_5Cl + HCl$$

Because chlorine is electronegative, PhCl exhibits decreased susceptibility to attack by other electrophiles. For this reason, the chlorination process produces only small amounts of dichloro and trichlorobenzenes.

Chloroform

Chloroform, also known as trichloromethane and methyl trichloride, is a chemical compound with formula $CHCl_3$. It does not support combustion in air, although it will burn when mixed with more flammable substances. It is a member of a group of compounds known as trihalomethanes. Chloroform has myriad uses as a reagent and a solvent. It is also considered an environmental hazard.

In 1847, the Edinburgh obstetrician James Young Simpson first used chloroform for general anesthesia during childbirth. The use of chloroform during surgery expanded rapidly thereafter in Europe. In the United States, chloroform began to replace ether as an anesthetic at the beginning of the 20th century; however, it was quickly abandoned in favour of ether upon discovery of its toxicity, especially its tendency to cause fatal cardiac arrhythmia analogous to what is now termed "sudden sniffer's death". Ether is still the preferred anesthetic in some developing nations due to its high therapeutic index and low price. Trichloroethylene, a halogenated aliphatic hydrocarbon related to chloroform, was proposed as a safer alternative, though it, too, was later found to be carcinogenic.

Production

Industrially, chloroform is produced by heating a mixture of chlorine and either chloromethane or methane. At 400-500°C, a free radical halogenation occurs, converting the methane or chloromethane to progressively more chlorinated compounds.

$$CH_4 + Cl_2 \longrightarrow CH_3Cl + HCl$$

$$CH_3Cl + Cl_2 \longrightarrow CH_2Cl_2 + HCl$$

$$CH_2Cl_2 + Cl_2 \longrightarrow CHCl_3 + HCl$$

Chloroform undergoes further chlorination to give CCl_4:

$$CHCl_3 + Cl_2 \longrightarrow CCl_4 + HCl$$

The output of this process is a mixture of the four chloromethanes: chloromethane, dichloromethane, chloroform (trichloromethane), and carbon tetrachloride, which are then separated by distillation.

Chloroform was first produced industrially by the reaction

of acetone (or ethanol) with sodium hypochlorite or calcium hypochlorite, known as the haloform reaction. The chloroform can be removed from the attendant acetate salts (or formate salts if ethanol is the starting material) by distillation.

This reaction is still used for the production of bromoform and iodoform. The haloform process is obsolete for the production of ordinary chloroform. It is, however, used to produce deuterated material industrially. Deuterochloroform is prepared by the reaction of sodium deuteroxide with chloral hydrate. All of the aldehyde hydrogen is retained in the product, though, and samples of higher isotopic purity are obtained from trichloroacetophenone as starting material.

Inadvertent Synthesis of Chloroform

Haloform-like reactions also occur inadvertently even in domestic settings. Sodium hypochlorite solution (chlorine bleach) and methyl ethyl ketone (nail-varnish remover) produces chloroform.

Pool chlorine and acetone will also produce chloroform.

Uses

The major use of chloroform today is in the production of the freon refrigerant R-22. However, as the Montreal Protocol takes effect, this use can be expected to decline as R-22 is replaced by refrigerants that are less liable to result in ozone depletion. In addition, it is used under research conditions to anesthetise mosquitoes for experiments, most frequently for the study of malaria.

As a Solvent

Chloroform is a common solvent because it is relatively unreactive, miscible with most organic liquids, and conveniently volatile. Small amounts of chloroform are used as a solvent in the pharmaceutical industry and for producing dyes and pesticides. Chloroform is an effective solvent for alkaloids in their base form and thus plant material is commonly extracted with chloroform for pharmaceutical processing.

For example, it is commercially used to extract morphine from poppies, scopolamine from *Datura* plants. Chloroform containing deuterium (heavy hydrogen), $CDCl_3$, is a common solvent used in NMR spectroscopy. It can be used to bond pieces of acrylic glass (which is also known under the trade name 'Perspex').

As a Reagent in Organic Synthesis

As a reagent, chloroform serves as a source of the dichlorocarbene CCl_2 group. It reacts with aqueous sodium hydroxide (usually in the presence of a phase transfer catalyst) to produce dichlorocarbene, CCl_2. This reagent effects ortho-formylation of activated aromatic rings such as phenols, producing aryl aldehydes in a reaction known as the Reimer-Tiemann reaction. Alternatively the carbene can be trapped by an alkene to form a cyclopropane derivative.

Safety

As might be expected from its use as an anesthetic, inhaling chloroform vapours depresses the central nervous system. Breathing about 900 parts of chloroform per million parts air (900 parts per million) for a short time can cause dizziness, fatigue, and headache.

Chronic chloroform exposure may .cause damage to the liver (where chloroform is metabolised to phosgene) and to the kidneys, and some people develop sores when the skin is immersed in chloroform.

Animal studies have shown that miscarriages occur in rats and mice that have breathed air containing 30 to 300 ppm chloroform during pregnancy and also in rats that have ingested chloroform during pregnancy. Offspring of rats and mice that breathed chloroform during.pregnancy have a higher incidence of birth defects, and abnormal sperm have been found in male mice that have breathed air containing 400 ppm chloroform for a few days. The effect of chloroform on reproduction in humans is unknown.

Chloroform once appeared in toothpastes, cough syrups, ointments, and other pharmaceuticals, but it has been banned in consumer products in the United States since 1976.

The NTP's eleventh report on carcinogens implicates it as reasonably anticipated to be a human carcinogen, a designation equivalent to IARC class 2A. It has been most readily associated with hepatocellular carcinoma. Caution is mandated during its handling in order to minimise unnecessary exposure; safer alternatives, such as dichloromethane, have resulted in a substantial reduction of its use as a solvent.

During prolonged storage hazardous amounts of phosgene can accumulate in the presence of oxygen and ultraviolet light. To prevent accidents commercial material is stabilised with ethanol or amylene, but samples that have been recovered or dried no longer contain any stabiliser and caution must be taken with those. Suspicious bottles should be tested for phosgene. Filter paper strips, wetted with 5 per cent diphenylamine, 5 per cent dimethylaminobenzaldehyde, and then dried, turn yellow in phosgene vapour.

Commonly used in DNA extractions and generally in conjunction with phenol to form a biolayer with extraction buffer (tris, etc.). DNA will form in the supernatant while protein and non-soluble cell materials will precipitate between the buffer chloroform layers.

Decahydronaphthalene

Decahydronaphthalene (also known as decalin, or as bicyclo[4.4.0]decane), a bicyclic organic compound, is an industrial solvent. A colourless liquid with an aromatic odour, it is used as a solvent for many resins. It is the saturated analog of naphthalene and can be prepared from it by hydrogenation in a fused state in the presence of a catalyst. Decahydronaphthalene easily forms explosive organic peroxides upon storage in the presence of air.

It occurs in cis and trans forms. The trans form is energetically more stable because of less steric interactions.

1,1-Dichloroethane

1,1-Dichloroethane is a chlorinated hydrocarbon. It is a colourless oily liquid with a chloroform-like odour. It is not easily soluble in water, but miscible with most organic solvents.

Large volumes of 1,1-dichloroethane are manufactured, with annual production exceeding 1 million pounds in the United States. It is mainly used as a feedstock in chemical synthesis, chiefly of 1,1,1-trichloroethane. It is also used as a solvent for plastics, oils and fats, as a degreaser, as a fumigant in insecticide sprays, in halon fire extinguishers, and in cementing of rubber. It is used in manufacturing of high-vacuum resistant rubber and for extraction of temperature-sensitive substances. Thermal cracking at 400-500°C and 10 MPa yields vinyl chloride. In the past, 1,1-dichloroethane was used as a surgical inhalational anesthetic.

In the atmosphere, 1,1-dichloroethane decomposes with half-time of 62 days, chiefly by reaction of photolytically produced hydroxyl radicals.

1,2-Dichloroethane

The chemical compound 1,2-dichloroethane, commonly known by its old name of ethylene dichloride (EDC), is a chlorinated hydrocarbon, mainly used to produce vinyl chloride monomer (VCM, chloroethene), the major precursor for PVC production. 1,2-Dichloroethane is also used generally as an intermediate for other organic chemical compounds, and as a solvent.

In 1794, a group of four Dutch friends under the name of *Gezelschap der Hollandsche Scheikensleishen* (Society of Dutch Chemists) consisted of physician Jan Rudolph Deerman, merchant Adriaan Paets van Troopstwijkity, chemist Anthoni Lauwerenburgest and botanist Nicolaas Bondtitigutrud. They were the first to produce 1,2-dichloroethane from olefiant gas (oil-making gas, ethylene) and chlorine gas. Although the *Gezelschap* in practice didn't do much in-depth scientific research, they and their publications where highly regarded.

Part of that acknowledgement is that 1,2-dichloroethane has been called Dutch oil in old chemistry.

Chemistry

1,2-Dichloroethane has chemical formula $C_2H_4Cl_2$.

Cf. 1,1-Dichloroethane (ethylidene dichloride).

Production

1,2-dichloroethane is primarily produced by iron (III) chloride catalysed reaction of ethene (ethylene) and chlorine:

$$H_2C{=}CH_2 + Cl_2 \longrightarrow Cl\text{-}CH_2\text{-}CH_2\text{-}Cl$$

In subsequent reactions, notably to vinyl chloride (chloroethene), hydrogen chloride is formed and reused in a copper (II) chloride catalysed reaction, to also produce 1,2-dichloroethane from ethene and oxygen:

$$H_2C{=}CH_2 + 2\ HCl + \tfrac{1}{2}\ O_2 \longrightarrow Cl\text{-}CH_2\text{-}CH_2\text{-}Cl + H_2O$$

Uses

Vinyl Chloride Monomer (VCM) Production: With approximately 80 per cent of the world's consumption of 1,2-dichloroethane, the major application is in the production of vinyl chloride monomer (VCM, chloroethene), which is the precursor to polyvinyl chloride under the formation of hydrogen chloride:

$$Cl\text{–}CH_2\text{–}CH_2\text{–}Cl \longrightarrow H_2C{=}CH\text{–}Cl + HCl$$

The hydrogen chloride can be reused in the production process, in the formation of more 1,2-dichloroethane.

Other uses

As a good apolar aprotic solvent, 1,2-dichloroethane is used as degreaser and paint remover. As an active building reagent, it is used as intermediate various organic compounds. Historically, it was used as an anti-knock additive in leaded fuels.

Safety

It occurs as a toxic, corrosive, highly flammable, and possibly carcinogenic colourless liquid with a chloroform-like odour.

Dichloromethane

Dichloromethane (DCM) or methylene chloride is the chemical compound with the formula CH_2Cl_2. It is a colourless, volatile liquid with a moderately sweet aroma. It is widely used as a solvent, the general view being that it is one of the less harmful of the chlorocarbons, and it is miscible with most organic solvents.

Dichloromethane was first prepared in 1840 by the French chemist Henri Victor Regnault, who isolated it from a mixture of chloromethane and chlorine that had been exposed to sunlight.

Production

Industrially, dichloromethane is produced by reacting either methyl chloride or methane with chlorine gas at 400-500°C. At these temperatures, both methane and methyl chloride undergo a series of reactions producing progressively more chlorinated products.

$$CH_4 + Cl_2 \longrightarrow CH_3Cl + HCl$$

$$CH_3Cl + Cl_2 \longrightarrow CH_2Cl_2 + HCl$$

$$CH_2Cl_2 + Cl_2 \longrightarrow CHCl_3 + HCl$$

$$CHCl_3 + Cl_2 \longrightarrow CCl_4 + HCl$$

The output of these processes is a mixture of methyl chloride, dichloromethane, chloroform, and carbon tetrachloride. These compounds are separated by distillation.

Uses

Dichloromethane's volatility and ability to dissolve a wide range of organic compounds makes it an ideal solvent for many chemical processes. It is widely used as a paint stripper and a degreaser. In the food industry, it is used to decaffeinate coffee and to prepare extracts of hops and other flavourings. It is used to chemically weld certain plastics (for example, it is used to seal the casing of electricmeters). Its volatility has led to its use as an aerosol spray propellant and as a blowing agent for polyurethane foams. It is also used as a fumigant pesticide for stored strawberries and grains. It is also the most common "active" component in Drinking Birds. Concerns about its health effects have led to a search for alternatives in many of these applications.

It is used in Christmas lights called bubble lights, in a sealed vial which bubbles when the incandescent light bulb below it is lit. Wurlitzer also used it in their 1940's "bubbler" style jukeboxes.

The bubble tubes were up to 30 inches long and used resistors to provide the heat to boil the liquid in a small constricted chamber that had bits of rock and a special glass valve to concentrate the small bubbles into larger ones. It is still used today in their reproduction machines. Dichloromethane is quite often used as a farming tool in Eastern and Central America as a gene adaptation tool.

Toxicity

Dichloromethane is the least toxic of the simple chlorohydrocarbons, but it is not without its health risks as its high volatility makes it an acute inhalation hazard. Dichloromethane is also metabolised by the body to carbon monoxide potentially leading to carbon monoxide poisoning. Prolonged skin contact can result in the dichloromethane dissolving some of the fatty tissues in skin, resulting in skin irritation or chemical burns.

It may be carcinogenic, as it has been linked to cancer of the lungs, liver, and pancreas in laboratory animals. Dichloromethane is a mutagen/teratogen and crosses the placenta, causing fetal toxicity in women who are exposed to it during pregnancy. In animal experiments it was fetotoxic at doses that were maternally toxic but no teratogenic effects were seen.

In many countries products containing dichloromethane must carry labels warning of its health risks.

Diethyl ether

Diethyl ether, also known as ether and ethoxyethane, is a clear, colourless, and highly flammable liquid with a low boiling point and a characteristic smell. It is an isomer of butanol. Diethyl ether has the formula $CH_3—CH_2—O—CH_2—CH_3$. It is used as a common solvent and has been used as a general anesthetic. Ether is sparingly soluble in water (6.9 g/100 ml).

Alchemist Raymundus Lullus is credited with discovering the compound in 1275 AD, although there is no contemporary evidence of this. It was first synthesised in 1540 by Valerius Cordus, who called it "oil of sweet vitriol" (*oleum dulci vitrioli*)—the name was due to the fact that it was originally discovered by distilling a mixture of ethanol and sulphuric acid (then known as oil of vitriol)—and noted some of its medicinal properties. At about the same time, Theophrastus Bombastus von Hohenheim, better known as Paracelsus, discovered ether's analgesic properties. The name *ether* was given to the substance in 1730 by A.S.Frobenius.

Anesthetic Use

The American doctor Crawford Williamson Long, M.D., was the first surgeon to use it as a general anesthetic, on March 30, 1842. William T.G. Morton was previously credited with the first public demonstration of ether anesthesia on October 16, 1846 at the Ether Dome in Boston, Massachusetts, although Dr. Crawford Long is now known to have demonstrated its use publicly to other officials in Georgia. Ether was sometimes used in place of Chloroform because it had a wider margin of error—that is, a larger difference between the recommended dosage and a toxic overdose.

Today, ether is rarely used. The use of flammable ether

waned as non-flammable anesthetic agents such as halothane became available. Additionally, ether had many undesirable side effects, such as post-anesthetic nausea and vomiting. Modern anesthetic agents, such as Methyl propyl ether (Neothyl) and methoxyflurane (Penthrane) reduce these side effects.

Because of its high volatility, low ignition point, and tendency to form explosive peroxides, diethyl ether must be used with care in laboratory settings.

Ether may be used to anesthetise ticks before removing them from an animal or a person's body. The anesthesia relaxes the tick and prevents it from maintaining its mouthpart under the skin.

Recreational use

The anesthetic effects of ether have made it a recreational drug, although not a popular one. Diethyl ether is not as toxic as other solvents used as recreational drugs.

Ether, mixed with ethanol, was marketed in the 19th century as a cure-all and recreational drug, during one of Western society's temperance movements. At the time, it was considered improper for women to consume alcoholic beverages at social functions, and sometimes ether-containing drugs would be consumed instead. A cough medicine called Hoffmann's Drops was marketed at the time as one of these drugs, and contained both ether and alcohol in its capsules. Ether tends to be difficult to consume alone, and thus was often mixed with drugs like ethanol for recreational use.

Due to its immiscibility with water and the fact that non-polar organic compounds are highly soluble in it, ether is also used in the production of freebase cocaine.

Synthesis

Diethyl ether is typically prepared both in laboratories and on an industrial scale by the acid ether synthesis. Ethyl alcohol is mixed with a strong dehydrating acid, typically sulphuric acid, H_2SO_4. The acid releases hydrogen ions into solution at very high concentrations. Some of these ions, as H^+, will seek out the relatively electronegative oxygen atoms on the ethyl alcohol molecules, creating a form of protonated ethanol with a positive charge:

$$CH_3\text{–}CH_2\text{–}OH + H^+ \longrightarrow CH_3\text{–}CH_2\text{–}OH_2^+$$

This species is highly electron-deficient (electrophilic) and positively charged, so it will attract non-protonated ethanol molecules, which contain the nucleophilic hydroxyl group. The resulting attack will result in the production of diethyl ether and water:

$$CH_3\text{–}CH_2\text{–}OH_2^+ + CH_3\text{–}CH_2\text{–}OH \longrightarrow H_2O + H^+ + CH_3\text{–}CH_2\text{–}O\text{–}CH_2\text{–}CH_3$$

Because the alcohol group is not a strong nucleophile (being uncharged), this reaction goes to equilibrium. In order to produce significant amounts of diethyl ether, distillation must be used to remove the low-boiling ether and drive the reaction forward. When performing this synthesis, it is advisable to use electric heating rather than an open-flame burner, as this will reduce (but by no means eliminate) the risk of fire. Also, note that the addition of sulphuric acid to the ethyl alcohol will release a great deal of heat, especially if the ethyl alcohol contains traces of water. Performing this addition carefully in an ice bath will prevent ether from boiling out prematurely.

Precautions

Diethyl ether is extremely flammable. Its vapours are denser than air and will accumulate if proper ventilation is not present. Simple static electricity will ignite ether vapours. Diethyl ether vapours ignite explosively, and should only be used inside a fume hood.

Additionally, diethyl ether is prone to peroxide formation, and can form explosive diethyl ether peroxide. Ether peroxides are higher boiling and are contact explosives when dry. Ether should never be distilled to dryness, as the risk of explosion increases dramatically. Diethyl ether is typically supplied with

BHT (2,6-di-tert-butyl-4-methylphenol), which reduces the formation of peroxides. Bottles older than 3 months should be routinely tested for peroxides. An iron wire, releasing Fe ions catalysing the peroxide decomposition, was often added to bottles with diethyl ether as a preventive measure, however, Fe (III) ions also strongly enhance peroxide formation. Storage over NaOH precipitates the intermediate ether hydroperoxides.

As Diethyl ether is also an anesthetic, inhalation of significant quantities of fumes can also cause severe side effects and in large doses, unconsciousness.

Tributyl phosphate

Tributyl phosphate (TBP), n-tributyl phosphate, or tri-n-butyl phosphate, is an odourless liquid, colourless to pale yellow in appearance, with applications in industrial and nuclear chemistry. It is slightly flammable and moderately dangerous to humans. It is an ester of orthophosphoric acid and *n*-butanol. It is a very good solvent.

Production

Tributyl phosphate is manufactured by esterification of orthophosphoric acid with butyl alcohol.

This is a high volume chemical with production estimated at 3,000 – 5,000 tons worldwide.

Use

Industrial: TBP is a solvent and plasticiser for cellulose esters (e.g. nitrocellulose and cellulose acetate). It forms stable hydrophobic complexes with some metals; these complex are soluble in organic solvents and in supercritical CO_2.

The major uses of TBP in industry are as a component of aircraft hydraulic fluid and as a solvent for extraction and purification of rare earth metals from their ores.

TBP finds its use as a solvent in inks, synthetic resins, gums, adhesives (namely for veneer plywood) and herbicide and fungicide concentrates.

As it has no odour, together with a large amount of, e.g. isopropyl alcohol it finds use as anti-foaming agent in most detergent solutions, and in various emulsions, paints, and adhesives. It is also found as a defoamer in ethylene glycol-borax antifreeze solutions.

In oil-based lubricants addition of TBP increases the oil film strength. It is used also in mercerising liquids, where it improves their wetting properties.

It is also used as a heat exchange medium.

Consumer Products

TBP is used in some consumer products such as herbicides and water thinned paints and tinting bases.

Nuclear Chemistry

A 15-40 per cent (usually about 30 per cent) solution of tributyl phosphate in kerosene or dodecane is used in the liquid-liquid extraction (solvent extraction) of uranium, plutonium and thorium from spent uranium nuclear fuel rods dissolved in nitric acid, as part of a nuclear reprocessing process known as PUREX. Because of this, the shipment of 20 tons of tributyl phosphate to North Korea from China in 2002, coinciding with the resumption of activity at Yongbyon Nuclear Scientific Research Centre, was seen by the United States and the International Atomic Energy Agency as cause for concern; that amount was considered sufficient to extract enough material for perhaps three to five potential nuclear weapons.

Hazards

The material will burn, but (in the absence of significant vaporisation) should not pose a particular explosive hazard. Inhalation and ingestion should be avoided due to possible central nervous system effects. A lab coat and safety glasses should be worn; a tributyl phosphate is not presently known to be, or suspected of being, a carcinogen, but may be mutagenic or have reproductive effects; consult the substance's MSDS for full details.

In contact with concentrated nitric acid the TBP-kerosene solution forms hazardous and explosive red oil.

Trihalomethane

Trihalomethanes (THMs) are chemical compounds in which three of the four hydrogen atoms of methane (CH_4) are replaced by halogen atoms. Many trihalomethanes find uses in industry as solvents or refrigerants. THMs are also environmental pollutants, and many are considered carcinogenic. Trihalomethanes with all the same halogen atoms are called haloforms.

Industrial uses

Refrigerants: Trifluoromethane and chlorodifluoromethane are both used as refrigerants in some applications. Trihalomethanes released to the environment break down faster than chlorofluorocarbons (CFCs), thereby doing much less damage to the ozone layer (if they contain chlorine). Chlorodifluoromethane is a refrigerant HCFC, or hydrochlorofluorocarbon, while fluoroform is an HFC, or hydrofluorocarbon. Fluoroform is not ozone depleting.

Unfortunately, the breakdown of trihalomethane HCFCs does still result in the creation of some free chlorine radicals in the upper atmosphere and subsequent ozone destruction. Ideally, HCFCs will be phased out entirely in favour of entirely non-chlorinated refrigerants.

Solvents

Chloroform (trichloromethane) is a very common solvent used in organic chemistry. It is a significantly less polar solvent than water, well-suited to dissolving many organic compounds.

Although still toxic and potentially carcinogenic, chloroform is significantly less harmful than carbon tetrachloride. Because of the health and regulatory issues associated with the use of carbon tetrachloride, in modern chemistry laboratories chloroform is used as a cheaper, cleaner alternative wherever possible.

Water Pollutants

Trihalomethanes are formed as a by-product when chlorine or bromine are used to disinfect water for drinking. They result from the reaction of chlorine and/or bromine with organic matter in the water being treated.

The THMs produced may have adverse health effects at high concentrations, and many governments set limits on the amount permissible in drinking water. In the United States, the EPA limits the total concentration of chloroform, bromoform, bromodichloro-methane, and dibromochloromethane to 80 parts per billion in treated water. This number is called "total trihalomethanes" (TTHM).

Chloroform is also formed in swimming pools which are disinfected with chlorine or hypochlorite in the haloform reaction with organic substances (urine, sweat and skin particles). The reaction to phosgene under the influence of UV is also possible.

Turpentine

Turpentine (also called spirit of turpentine, oil of turpentine, wood turpentine, gum turpentine) is a fluid obtained by the distillation of resin obtained from trees, mainly pine trees. It is composed of terpenes, mainly the monoterpenes alpha-pinene and beta-pinene.

It has a potent odour similar to that of nail polish remover. It is sometimes known colloquially as *turps*, but this more often refers to turpentine substitute (or mineral turpentine).

The word *turpentine* is formed (via French and Latin) from the Greek word *terebinthine,* the name of a species of tree, the terebinth tree, from whose sap the spirit was originally distilled.

Production

One of the earliest sources was the terebinth or turpentine tree (*Pistacia terebinthus*), a Mediterranean tree related to the pistachio.

Important pines for turpentine production include:

- Maritime Pine: *Pinus pinaster*
- Aleppo Pine: *Pinus halepensis*
- Masson's Pine: *Pinus massoniana*
- Sumatran Pine: *Pinus merkusii*
- Longleaf Pine: *Pinus palustris*
- Loblolly Pine: *Pinus taeda*
- Ponderosa Pine: *Pinus ponderosa*

Industrial Uses

The two primary uses of turpentine in industry are as a solvent and as a source of materials for organic synthesis.

As a solvent, turpentine is used for thinning oil-based paints, for producing varnishes, and as a raw material for the chemical industry.

Its industrial use as a solvent in industrialised nations has largely been replaced by the much cheaper turpentine substitutes distilled from crude oil.

Canada balsam, also called Canada turpentine or balsam of fir, is a turpentine which is made from the resin of the balsam fir.

Venice turpentine is produced from the Western Larch *Larix occidentalis*.

Turpentine is also used as a source of raw materials in the synthesis of fragrant chemical compounds. Commercially used camphor, linalool, alpha-terpineol, and geraniol are all usually produced from alpha-pinene and beta-pinene, which are two of the chief chemical components of turpentine. These pinenes are separated and purified by distillation. The mixture of diterpenes and triterpenes that is left as residue after turpentine distillation is sold as rosin.

Turpentine is also added to many cleaning and sanitary products due to its antiseptic properties and its "clean scent".

Medicinal uses

Turpentine has been used medically since ancient times.

- Applied externally to the affected areas, turpentine is a highly effective treatment for lice.
- Turpentine can be mixed with animal fat as a primitive chest rub for nasal and throat ailments. Many modern chest rubs still contain some turpentine (e.g., Vicks).
- Internal administration of turpentine is no longer common today, though it was once the preferred means of treating intestinal parasites due to its antiseptic and diuretic properties.
- Drinking turpentine is extremely dangerous and can be life threatening. In addition, drinking turpentine is not an effective way to induce an abortion

Hazards

Turpentine is an organic solvent, and thus poses many of the same hazards as do other such substances. Its vapour can burn the skin and eyes, damage the lungs and respiratory system, as well as the central nervous system when inhaled, and cause renal failure when ingested, among other things. It is highly flammable.

Sulpholane

Sulpholane (also tetramethylene sulphone, systematic name: 2,3,4,5-tetrahydrothiophene-1,1-dioxide) is a clear, colourless liquid commonly used in the chemical industry as an extractive distillation solvent or reaction solvent. Sulpholane was originally developed by the Shell Oil Company in the 1960s as a solvent to purify butadiene. Sulpholane is an aprotic organosulphur compound, and it is readily soluble in water.

Chemical Properties

Sulpholane is classified as a sulphone, a group of organosulphur compounds containing a sulphonyl functional group. The sulphonyl group is a sulphur atom doubly bonded to two oxygen atoms. The sulphur-oxygen double bond is highly polar, allowing for its high solubility in water, while the four carbon ring provides non-polar stability.

It is these properties that allow it to be so miscible in both water and hydrocarbons, resulting in its widespread use as a solvent for purifying hydrocarbon mixtures.

Synthesis

The original method developed by the Shell Oil Company was to first allow butadiene to react with sulphur dioxide. This yields sulpholene, which was then hydrogenated using Raney nickel as a catalyst to give sulpholane.

O=S=O + butadiene ⟶ sulpholene + H_2 ⟶ sulpholane

Shortly thereafter, it was discovered that both the product yield and the lifetime of the catalyst could be improved by adding hydrogen peroxide and then neutralising to a pH of roughly 5-8 before hydrogenation.

Developments have continued over the years, including in the catalysts used. Recently, it was found that Ni-B/MgO showed superior catalytic activity to that of Raney nickel and other common catalysts that have been used in the hydrogenation of sulpholene.

Other syntheses have also been developed, such as oxidising tetrahydrothiophene with hydrogen peroxide. This first produces tetramethylene sulphoxide, which can then be further oxidised to tetramethylene sulphone.

Because the first oxidation takes place at low temperature and the second at relatively higher temperature, the reaction can be controlled at each stage.

This gives greater freedom for the manipulation of the reaction, which can potentially lead to higher yields and purity.

Uses

Sulpholane is widely used as an industrial solvent, especially in the extraction of aromatic hydrocarbons from hydrocarbon mixtures and to purify natural gas.

The first large scale commercial use of sulpholane, the sulphinol process, was first implemented by Shell Oil Company in March of 1964 at the Person gas plant near Karnes City, Texas. The sulphinol process purifies natural gas by removing H_2S, CO_2, COS and mercaptans from natural gas with a mixture of alkanolamine and sulpholane.

Shortly after the sulphinol process was implemented, sulpholane was found to be highly effective in separating high purity aromatic compounds from hydrocarbon mixtures using liquid-liquid extraction. This process is still widely used today in refineries and the petrochemical industry. Because sulpholane is the most efficient industrial solvent for purifying aromatics, they operate at the lowest solvent-to-feed ration, making sulpholane units highly cost effective. In addition, it is selective in a range that complements distillation; where sulpholane can't separate two compounds, distillation easily can and vice versa, keeping sulpholane units useful for a wide range of compounds with minimal additional cost.

While sulpholane is highly stable and can therefore be reused many times, it does eventually break down into acidic by-products. A number of measures have been developed to remove these by-products, allowing the sulpholane to be reused and increase the lifetime of a given supply. Some methods that have been developed to regenerate spent sulpholane include vacuum and steam distillation, back extraction, adsorption, and anion-cation exchange resin columns.

Morpholine

Morpholine is an organic chemical compound having the chemical formula $O(CH_2CH_2)_2NH$. This heterocycle, pictured at right, features both amine and ether functional groups. Because of the amine, morpholine is a base; its conjugate acid is called morpholinium. For example, when morpholine is neutralised by hydrochloric acid, one obtains the salt morpholinium chloride.

Uses

Industrial Applications: Morpholine is a common additive, in ppm concentrations, for pH adjustment in both fossil fuel and nuclear power plant steam systems. Morpholine is used because its volatility is about the same as water, so once it is added to the water, its concentration becomes distributed rather evenly in both the water and steam phases. Its pH adjusting qualities then become distributed throughout the steam plant to provide corrosion protection.

Morpholine is often used in conjunction with low concentrations of hydrazine or ammonia to provide a comprehensive all-volatile treatment chemistry for corrosion protection for the steam systems of such plants. Morpholine decomposes reasonably slowly in the absence of oxygen even at the high temperatures and pressures in these steam systems.

Organic Synthesis

Morpholine undergoes most chemical reactions typical for other secondary amines, though the presence of the ether oxygen withdraws electron density from the nitrogen, rendering it less nucleophilic (and less basic) than structurally similar secondary amines such as piperidine. For this reason, it forms a stable chloramine.

It is commonly used to generate enamines. Morpholine is widely used in organic synthesis. For example, it is a building block in the preparation of the antibiotic linezolid and the anti-cancer agent gefitinib.

Morpholine is used as a chemical emulsifier in the process of waxing fruit. Fruits make waxes naturally to protect against insects and fungal contamination, but this can be lost by means of the food processing companies when they clean the fruit. As a result, an extremely small amount of new wax is applied and morpholine is then added and used as an emulsifier to evenly coat a fruit with the wax.

In research and in industry, the low cost and polarity of morpholine lead to its common use as a solvent for chemical reactions.

As a Component in Fungicides

Morpholine derivatives used as agricultural fungicides in cereals are known as Ergosterol Biosynthesis Inhibitors.

- Morpholines
  - — Fenpropimorph
  - — Tridemorph
- Piperidines
  - — Fenpropidin

Mineral Spirits

Mineral Spirits is a petroleum distilate commonly used as a paint thinner and mild solvent. In Europe, it is referred to as petroleum spirit or white spirit. In industry, mineral spirits is used for cleaning and degreasing machine tools and parts.

According to Wesco, a supplier of solvents and cleaning equipment, mineral spirits are especially effective in removing oils, greases, carbon, and other material from metal. Mineral spirits may also be used in conjunction with cutting oil as a thread cutting and reaming lubricant.

Artists use mineral spirits as an alternative to turpentine, one that is both less flammable and less toxic. Because of interactions with pigments, artists require a higher grade of mineral spirits than many industrial users, including the complete absence of residual sulphur. Odourless Mineral Spirits are mineral spirits that have been further refined to remove the more toxic aromatic compounds, and are recommended for applications such as oil painting, where humans have close contact with the solvent. According to Robert Gamblin of Gamblin Artists Colours, the best brands contain 100 per cent aliphatic compounds, and dry without leaving any residue. A typical composition for mineral spirits is the following: aliphatic solvent hexane having a maximum aromatic hydrocarbon content of 0.1 per cent by volume, a kauri-butanol value of 29, an initial boiling point of 149°F (65°C), a dry point of approximately 156°F (69°C), and a specific mass of 0.7 g/cc.

1,4-Dioxane

1,4-Dioxane, often just called dioxane, is a clear, colourless heterocyclic organic compound which is a liquid at room temperature and pressure. It has the molecular formula $C_4H_8O_2$ and a boiling point of 101°C. It is commonly used as an aprotic solvent. 1,4-Dioxane has a weak smell similar to that of diethyl ether. There are also two other less common isomeric compounds, 1,2-dioxane and 1,3-dioxane. 1,2-Dioxane is a peroxide which forms naturally in old bottles of tetrahydrofuran.

1,4-Dioxane is classified as an ether, with each of its two oxygen atoms forming an ether functional group. It is more polar than diethyl ether, which also has four carbons, but only one ether functional group.

Diethyl ether is rather insoluble in water, but 1,4-dioxane is miscible with water and is hygroscopic. Its higher polarity and slightly higher molecular mass also gives it a substantially higher boiling point than diethyl ether. When used as a solvent for a grignard reaction, Dioxane favourably affects the formation of magnesium halide salts in the Schlenk equilibrium.

The name *dioxane* should not be confused with *dioxin,* which is a different compound but is also a diether (two ether functional groups).

Uses

1,4-Dioxane is primarily used in solvent applications for the manufacturing sector; however, it is also found in fumigants and automotive coolant. Additionally, the chemical is also used as a foaming agent and appears as an accidental by-product of the ethoxylation process in cosmetics manufacturing. It may contaminate cosmetics and personal care products such as deodorants, shampoos, toothpastes and mouthwashes.

Safety Concerns

Like many other ethers, dioxanes combine with atmospheric oxygen on standing to form explosive peroxides, similar to many other ethers. Distillation of dioxanes concentrates these peroxides increasing the danger. Appropriate precautions should be taken.

1,4-dioxane is a known eye and respiratory tract irritant. It is suspected of causing damage to the central nervous system, liver and kidneys. Accidental worker exposure to 1,4-dioxane has resulted in several deaths. Dioxane is classified by the IARC as a Group 2B carcinogen: *possibly carcinogenic to humans* due to the fact that it is a known carcinogen in animals.

Ethyl Acetate

Ethyl acetate is the organic compound with the formula $CH_3CH_2OC(O)CH_3$. This colourless liquid has a characteristic, not unpleasant smell (similar to pear drops) like certain glues or nail polish removers, in which it is used. As the ester derived from ethanol and acetic acid, EtOAc, as it is commonly abbreviated, is manufactured on a large scale for use as a solvent.

Properties

Ethyl acetate is a moderately polar solvent that has the advantages of being volatile, relatively non-toxic, and non-hygroscopic. It is a weak hydrogen bond acceptor, and is not a donor due to the lack of an acidic proton (one directly bonded to an electronegative atom such as fluorine, oxygen, or nitrogen). Ethyl acetate can dissolve up to 3 per cent water and has a solubility of 8 per cent in water at room temperature. At elevated temperature its solubility in water is higher. It is unstable in the presence of strong aqueous bases and acids.

Uses and Occurrence

Ethyl acetate is widely employed as a solvent for nail varnishes and nail varnish removers. Industrially it is used to decaffeinate coffee beans and tea leaves. In chemistry, it is often mixed with a non-polar solvent such as hexanes as a chromatography solvent. It is also used as a solvent for extractions. It is rarely used as a reaction mixture due to the weakness of the ester linkage — it hydrolyses in the presence of strong acids and bases to give acetic acid and ethanol.

Ethyl acetate is also present in confectionery, perfumes, and fruits. It is used in perfumes because it evaporates at a fast rate, leaving but the scent of the perfume on the skin. It also confers a fruity smell, as do most esters. It is also used in paints as an activator or hardener.

Occurrence in Wines

Ethyl acetate is present in wines. It may be considered a contaminant at too high concentrations, as typically occurs when wine is exposed to air for a prolonged period. When present at too high concentration in wine, it is regarded as an off-flavour.

Other uses

In the field of entomology, ethyl acetate is an effective poison for use in insect collecting and study. In a killing jar charged with ethyl acetate, the vapours will kill the collected (usually adult) insect quickly without destroying it. Because it is not hygroscopic, ethyl acetate also keeps the insect soft enough to allow proper mounting suitable for a collection.

Synthesis

Ethyl acetate is synthesised via the Fischer esterification reaction from acetic acid and ethanol, typically in the presence of an acid catalyst such as sulphuric acid.

$$CH_3CH_2OH + CH_3COOH \longrightarrow CH_3COOCH_2CH_3 + H_2O$$

Because the reaction is reversible and produces an

equilibrium, the yield is low unless water is removed. In the laboratory, the ethyl acetate product can be isolated from water using a Dean-Stark apparatus.

Reactions

Ethyl acetate can be hydrolysed in acid or basic conditions to regain acetic acid and ethanol. The use of an acid catalyst such as sulphuric acid gives poor yields due it being an equilibrium — the reverse reaction of the Fischer esterification.

To obtain high yields, it is preferable to use a stoichiometric amount of strong base, such as sodium hydroxide. This reaction gives ethanol and sodium acetate, which is not able to react with ethanol any longer:

$$CH_3CO_2C_2H_5 + NaOH \longrightarrow C_2H_5OH + CH_3CO_2Na$$

Ethylene Glycol

Ethylene glycol (monoethylene glycol (MEG), IUPAC name: ethane-1,2-diol) is an alcohol with two -OH groups (a diol), a chemical compound widely used as an automotive antifreeze. In its pure form, it is an odourless, colourless, syrupy liquid with a sweet taste. Ethylene glycol is toxic, and its accidental ingestion should be considered a medical emergency.

Ethylene glycol was first prepared in 1859 by the French chemist Charles-Adolphe Wurtz. It was produced on a small scale during World War I as a coolant and as an ingredient in explosives. Widespread industrial production began in 1937 when ethylene oxide, a component in its synthesis, became cheaply available.

When first introduced it created a minor revolution in aircraft design because when used in place of water as an engine coolant, its higher boiling point allowed for smaller radiators operating at higher temperatures. Prior to the widespread availability of ethylene glycol, many aircraft manufacturers tried to use evaporative cooling systems which used water at high pressure. Invariably, these proved to be rather unreliable and were easily damaged in combat because they took up large amounts of room on the plane, where they were easily hit by gunfire.

Production

Ethylene glycol is produced from ethylene, via the intermediate ethylene oxide. Ethylene oxide reacts with water to produce ethylene glycol according to the chemical equation:

$$C_2H_4O + H_2O \longrightarrow HOCH_2CH_2OH$$

This reaction can be catalysed by either acids or bases, or can occur at neutral pH under elevated temperatures. The highest yields of ethylene glycol occur at acidic or neutral pH with a large excess of water. Under these conditions, ethylene glycol yields of 90 per cent can be achieved. The major by-products are the ethylene glycol oligomers diethylene glycol, triethylene glycol, and tetraethylene glycol.

Uses

The major use of ethylene glycol is as a coolant or antifreeze in, for example, automobiles and personal computers. Due to its low freezing point, it is also used as a deicing fluid for windshields and aircraft. Ethylene glycol has become increasingly important in the plastics industry for the manufacture of polyester fibres and resins, including polyethylene terephthalate, which is used to make plastic bottles for soft drinks. The antifreeze capabilities of ethylene glycol have made it an important component of vitrification mixtures for low-temperature preservation of biological tissues and organs.

Minor uses of ethylene glycol include the manufacture of capacitors, as a chemical intermediate in the manufacture of 1,4-dioxane and as an additive to prevent the growth of algae in liquid cooling systems for personal computers.

Ethylene glycol's high boiling point and affinity for water makes it an ideal desiccant for natural gas production. In the field, excess water vapour is usually removed by glycol dehydration. Glycol flows down from the top of a tower and meets a rising mixture of water vapour and hydrocarbon gases from the bottom. The glycol chemically removes the water vapour, allowing dry gas to exit from the top of the tower. The glycol and water are separated, and the glycol cycles back through the tower.

Ethylene glycol is also used in the manufacture of some vaccines, but it is not itself present in these injections. It is used as a minor (1-2 per cent) ingredient in shoe polish and also in some inks and dyes.

Ethylene glycol is commonly used in laboratories to precipitate out proteins in solution. This is often an intermediary step in fractionation, purification and/or crystallisation. It can be used to protect functional groups from reacting during organic synthesis. To get the functional group back to its original composition, simply add water and acid.

Ethylene glycol has seen some use as a rot and fungal treatment for wood, both as a preventative and a treatment after the fact. It has been used in a few cases to treat partially rotted wooden objects to be displayed in museums. It is one of only a few treatments that are successful in dealing with rot in wooden boats, and is relatively cheap.

Ethylene glycol is commonly used as a preservative for specimens in schools, frequently during dissection. It is said to be safer than formaldehyde, but the safety is questionable.

Toxicity

The major danger from ethylene glycol is following ingestion. Due to its sweet taste, children and animals will sometimes consume large quantities of it if given access to antifreeze. Ethylene glycol may also be found as a contaminant in unlawfully distilled whiskey (moonshine) made in a still constructed using an improperly washed car radiator. In developed countries, a bittering agent called denatonium/ denatonium benzoate, is generally added to ethylene glycol preparations as an adversant (to prevent accidental ingestion). If one has ingested ethylene glycol, give the person an alcoholic beverage while the paramedics arrive. Alcohol acts as a competitive inhibitor to the active site of the enzyme that converts ethylene glycol to its toxic form. Once alcohol binds, the ethylene glycol is harmlessly excreted out of the body.

Ethylene glycol poisoning is a medical emergency and in all cases a poison control centre should be contacted or medical attention should be sought. It is highly toxic with an estimated LD_{100} in humans of approximately 1.4 ml/kg. However, as little as 30 millilitres (2 tablespoons) can be lethal to adults.

Symptoms

Symptoms of ethylene glycol poisoning usually follow a three-step progression, although poisoned individuals will not always develop each stage or follow a specific time frame. Stage 1 consists of neurological symptoms including victims appearing to be intoxicated, exhibiting symptoms such as dizziness, headaches, slurred speech, and confusion.

Over time, the body metabolises ethylene glycol into other toxins, it is first metabolised to glycolaldehyde, which is then oxidised to glycolic acid, glyoxylic acid, and finally oxalic acid. Stage 2 is a result of accumulation of these metabolites and consists of tachycardia, hypertension, hyperventilation, and metabolic acidosis. Stage 3 of ethylene glycol poisoning is the result of kidney injury, leading to acute kidney failure. Oxalic acid reacts with calcium and forms calcium oxalate crystals in the kidney.

Treatment

Initial treatment consists of stabilising the patient and gastric decontamination. As ethylene glycol is rapidly absorbed, gastric decontamination needs to be performed soon after ingestion to be of benefit. Gastric lavage or nasogastric aspiration of gastric contents are the most common methods employed in ethylene glycol poisoning. Ipecac induced emesis or activated charcoal (charcoal does not adsorb glycols) are not recommended.

The antidotes for ethylene glycol poisoning are ethanol or

fomepizole; antidotal treatment forms the mainstay of management following ingestion. Ethanol (usually given IV as a 5 or 10 per cent solution in 5 per cent dextrose and water, but, also sometimes given in the form of a strong spirit such as whisky, vodka or gin) acts by competing with ethylene glycol for the enzyme alcohol dehydrogenase thus limiting the formation of toxic metabolites. Fomepizole acts by inhibiting alcohol dehydrogenase, thus blocking the formation of the toxic metabolites.

In addition to antidotes, hemodialysis can also be used to enhance the removal of unmetabolised ethylene glycol, as well as its metabolites from the body. Hemodialysis also has the added benefit of correcting other metabolic derangements or supporting deteriorating kidney function caused by ethylene glycol ingestion. Often both antidotal treatment and hemodialysis are used together in the treatment of poisoning.

Industrial Hazards

Ethylene glycol can begin to breakdown at 230°-250°F. Note that breakdown can occur when the system bulk (average) temperature is below these limits because surface temperatures in heat exchangers and boilers can be locally well above these temperatures.

The electrolysis of ethylene glycol solutions with a silver anode results in an exothermic reaction. The Apollo 1 fire catastrophe was caused by this reaction. The ethylene glycol–water mixture was ignited and was able to burn in the atmosphere of pure low pressure oxygen.

Carbonyl Chemistry

Ethylene glycol may also be used as a protecting group for carbonyls during synthesis. Acid catalysis, and a ketone or aldehyde with ethylene glycol will form a cyclic structure at the carbonyl. Other chemistry can then be done to the molecule before more acid will break open the protecting ring and restore the carbonyl.

Furfural

The chemical compound furfural is an industrial chemical derived from a variety of agricultural by-products, including corncobs, oat and wheat bran, and sawdust. The name *furfural* comes from the Latin word *furfur*, meaning bran, referring to its usual source.

Furfural is an aromatic aldehyde, with the ring structure shown at right. Its cl .mical formula is $C_5H_4O_2$. In its pure state, it is a colourless oily liquid with the odour of almonds, but upon exposure to air it quickly becomes yellow.

Furfural was first isolated in 1832 by the German chemist Johann Wolfgang Dobereiner, who formed a very small quantity of it as a by-product of formic acid synthesis. At the time, formic acid was formed by the distillation of dead ants, and Dobereiner's ant bodies probably contained some plant matter.

In 1840, the Scottish chemist John Stenhouse found that the same chemical could be produced by distilling a wide variety of crop materials, including corn, oats, bran, and sawdust, with aqueous sulphuric acid, and he determined that this chemical had an empirical formula of $C_5H_4O_2$. In 1901, the German chemist Carl Harries deduced furfural's structure.

Except for occasional use in perfume, furfural remained a relatively obscure chemical until 1922, when the Quaker Oats Company began mass-producing it from oat hulls. Today, furfural is still produced from agricultural by-products like sugarcane bagasse and corn cobs.

Properties

Furfural dissolves readily in most polar organic solvents, but is only slightly soluble in either water or alkanes.

Chemically, furfural participates in the same kinds of reactions as other aldehydes and other aromatic compounds. The aromatic stability of furfural is not as great as in benzene, and furfural participates in hydrogenation and other addition reactions more readily than many other aromatics.

When heated above 250°C, furfural decomposes into furan and carbon monoxide, sometimes explosively.

When heated in the presence of acids, furfural irreversibly solidifies into a hard thermosetting resin.

Production

Many plant materials contain the polysaccharide hemicellulose, a polymer of sugars containing five carbon atoms each. When heated with sulphuric acid, hemicellulose undergoes hydrolysis to yield these sugars, principally xylose. Under the same conditions of heat and acid, xylose and other five carbon sugars undergo dehydration, losing three water molecules to become furfural:

$$C_5H_{10}O_5 \longrightarrow C_5H_4O_2 + 3\ H_2O$$

For crop residue feedstocks, about 10 per cent of the mass of the original plant matter can be recovered as furfural. Furfural and water evaporate together from the reaction mixture, and separate upon condensation.

Global total capacity of production is about 450,000 tons. China is the biggest supplier of this product and they have about a half of global capacity. In the laboratory, synthesis of furfural from corn cobs takes place by reflux with dilute sulphuric acid.

Uses

Furfural is used as a solvent in petrochemical refining to extract dienes (which are used to make synthetic rubber) from other hydrocarbons.

Furfural, as well as its derivative furfuryl alcohol, can be used either by themselves or in together with phenol, acetone, or urea to make solid resins. Such resins are used in making fibreglass, some aircraft components, and automotive brakes.

Furfural is also used as a chemical intermediate in the production of the solvents furan and tetrahydrofuran. Hydroxymethylfurfural has been identified in a wide variety of heat processed foods.

Safety

When ingested or inhaled, furfural can cause symptoms similar to those of intoxication, including euphoria, headache, dizziness, nausea, and eventual unconsciousness and death due to respiratory failure. Contact with furfural irritates the skin and respiratory tract and can cause the lungs to fill with fluid.

Chronic skin exposure can lead to a skin allergy to the substance, as well as an unusual susceptibility to sunburn. In toxicity studies, furfural has led to tumours, mutations, and liver and kidney damage in animals.

Methyl tert-butyl Ether

Methyl tertiary-butyl ether (MTBE) is a chemical compound with molecular formula $C_5H_{12}O$. MTBE is a volatile, flammable and colourless liquid that is highly soluble in water. MTBE has a minty odour vaguely reminiscent of diethyl ether, leading to unpleasant taste and odour in water. MTBE is used in organic chemistry as a relatively inexpensive solvent with properties comparable to diethyl ether but with a higher boiling point and lower solubility in water. It is also used medically to dissolve gallstones.

Production

MTBE is manufactured by the chemical reaction of methanol and isobutylene. It was produced in very large quantities (more than 200,000 barrels per day in the United States in 1999) when it was being used widely as a fuel additive.

Because of widespread releases of MTBE-containing gasoline from Underground Storage Tanks all over the US, various jurisdictions banned the use of MTBE and production was reduced. MTBE contamination in drinking water aquifers is a serious concern in many states (most famous cases are Lake Tahoe and Santa Monica). By late 2006, most American gasoline retailers had ceased using MTBE as an oxygenate, and accordingly, US production had declined.

Physical Properties

MTBE forms azeotropes with water (52.6°C) and methanol (51.3°C).

Uses

MTBE is almost exclusively used as a fuel component in motor gasoline. It is one of a group of chemicals commonly known as oxygenates because they raise the oxygen content of gasoline.

As Anti-knocking Agent

In the US it has been used in gasoline at low levels since 1979 to replace tetra-ethyl lead to increase its octane rating and help prevent engine knocking. Oxygen helps gasoline burn more completely, reducing tailpipe emissions from pre-1984 motor vehicles.

In more modern vehicles, the emissions reduction is negligible. In one respect, the oxygen dilutes or displaces gasoline components such as aromatics (e.g. benzene) and sulphur. In another, oxygen optimises the oxidation during combustion. Most refiners have chosen to use MTBE over other oxygenates primarily for its blending characteristics and for economic reasons. It is produced from natural gas, which is less expensive than oil.

Since 1992, MTBE has been used at higher concentrations in some gasoline to fulfil the oxygenate requirements set by Congress in Clean Air Act amendments; however, since 1999, in California and other locations MTBE has begun to be phased out because of groundwater contamination (California Air Resources Board, 2004), citing unproven health effects. Due to its higher solubility in water MTBE moves more quickly than other fuel components (California Air Resources Board, 2004). The Energy Policy Act of 2005 reduces the federal requirement for oxygen content in reformulated gasoline.

In 1995, high levels of MTBE were unexpectedly discovered in the water wells of Santa Monica, California, and the US

Geological Survey reported detections. Subsequent US findings indicate tens of thousands of contaminated sites in water wells distributed across the country. As per toxicity alone, MTBE is not classified as a hazard for the environment. The maximum contaminant level of MTBE in drinking water has not yet been established by the EPA.

The leakage problem is partially attributed to the lack of effective regulations for underground storage tanks, but spillage from overfilling remains an important upset scenario. As an ingredient in unleaded gasoline, MTBE is the most soluble part. When dissolved in groundwater, MTBE will lead the contaminant plume with the remaining components such as Benzene, Toluene, etc. to follow. Thus the discovery of MTBE in public groundwater wells indicates that the contaminant source was a gasoline release. The MTBE concentrations used in the EU (usually 1.0-1.6 per cent) and allowed (maximum 5 per cent) in Europe are lower than in California.

Alternatives

Other compounds are available as oxygenate additives for gasoline, for example ethanol and related ethers, e.g. tert-amyl methyl ether (TAME). Reasons for using MTBE include economic considerations, as some of the production is obtained by adding methanol to isobutylene produced as a by-product of other processes. However, most MTBE facilities have to manufacture the methanol and isobutylene required to produce MTBE.

Ethanol has been advertised as a safe alternative by the agricultural interest groups in the USA and Europe. Its lack of toxicity is not different from MTBE, but as a polar solvent, it drives off non-polar hydrocarbons from the gasoline, a problem that MTBE does not cause.

Volatile hydrocarbons from gasoline are known (and severe) carcinogens and the main contributor to photochemical smog. EU's agricultural subsidies have produced an oversupply of wine, and the excess low-quality wine is being refined to ethanol fuel in Europe. This gives rise to political motives for supporting ethanol over MTBE. However, the political stability of the supply is a major advantage for ethanol and other biofuels.

Advocates of both sides of the debate in the United States sometimes claim that gasoline manufacturers have been forced to add MTBE to gasoline by law. It might be more correct to say they have been induced to do so, although any oxygenate would fulfil the law.

In 2003, California was the first state to start replacing the MTBE with ethanol. Several other states started switching soon afterwards.

Higher quality gasoline is also an alternative, i.e. so that additives such as MTBE are unnecessary. Iso-octane itself is used. MTBE plants can be retrofitted to produce iso-octane from isobutylene, which is a lighter-than-gasoline hydrocarbon and thus more difficult to sell. Iso-octane is the ideal gasoline, being the standard reference for the octane rating.

In the long run, diesel fuel is also an alternative, although it requires a major switchover to diesel-run cars. There are several varieties of biodiesel; both oxygen-containing methyl ethers and no-oxygen alkyl biodiesels are available.

In Chemistry

Being an ether, MTBE is a Lewis base. However, unlike other ethers such as diethyl ether or THF, it does not coordinate well enough with magnesium to be used for making Grignard reagents. The *tert*-butyl group is easily cleaved off under strongly acidic conditions (forming a moderately stable carbocation), particularly if heated (isobutylene is lost), something which can limit the use of MTBE as a solvent. However it possesses one distinct advantage over most ethers-it has a much lower tendency to form explosive organic peroxides than most ethers. Opened bottles of diethyl ether or THF can build up dangerous levels of these peroxides in months, whereas samples of MTBE are usually safe for years (but they should still be tested periodically).

Persistence and Pervasiveness in the Environment

MTBE is often introduced into water-supply aquifers by leaking underground storage tanks (USTs) at gasoline stations. Although USTs are much better constructed now than in the 1980s, accidental releases still take place because of the very large number of USTs. The high solubility and persistence of MTBE cause it to travel faster and farther than many other gasoline components when released into an aquifer.

MTBE has widespread occurrences in the aquifers of North America, where the majority of groundwater chemistry data has been acquired. As one regional example, the San Francisco Bay Area Regional Water Quality Control Board has indicated MTBE is one of the groundwater pollutants of most widespread concern in this major metropolitan region of the USA.

Health Risks

The IARC, a cancer research agency of the World Health Organisation, maintains MTBE is not classifiable as a human carcinogen. However, exposure to large doses of MTBE has significant non-cancer-related health risks. MTBE ruins the taste of water at concentrations of 5-15 μg/litre so that significant concentrations of MTBE in drinking water are detectable.

MTBE is not classified as a human carcinogen in low exposure levels by the International Agency for Research on Cancer (IARC). However, it has been shown to cause kidney lesions in animals. As an ether, MTBE acts as an emulsifier, increasing the solubility of other, harmful components of gasoline (for example, the known carcinogen benzene). It thus may increase the risk of contamination by other compounds. MTBE is biodegraded very slowly, remaining in water for decades or more. In addition some MTBE degrades in blood to the associated alcohol, tert-butanol, with a greatly increased residence time. The prolonged presence of this alcohol derivative is not fully understood.

Some industry advocates of MTBE contend that it has little

provable effect on humans, although its manufacturers did not test it for its effects on human health before introducing it as an additive. As of 2007, researchers have limited data about the health effects from ingestion of MTBE. The Environmental Protection Agency (EPA) has concluded that available data are not adequate to quantify health risks of MTBE at low exposure levels in drinking water, but that the data support the conclusion that MTBE is a potential human carcinogen at high doses.

Legislation and Litigation

United States: Estimates of the cost of removing MTBE from groundwater and soil contamination range from $1 to $30 billion, including removing the compound from aquifers and municipal water supplies and replacing leaky underground oil tanks. Some controversy centres on who will pay the costs of this remediation. In one case, the cost to oil companies to clean-up the MTBE in wells belonging to Santa Monica is estimated to exceed $200 million.

Recent state laws have been passed to ban MTBE in certain areas. California and New York, which together accounted for 40 per cent of US MTBE consumption, banned the chemical starting January 1, 2004, and as of September, 2005, twenty-five states had signed legislation banning MTBE.

In 2000, the US EPA drafted plans to phase out the use of MTBE nationwide over four years. Upon taking office, the Bush administration cancelled those plans.

In April of 2002, a California jury found several oil companies guilty of irresponsibly manufacturing and distributing MTBE, stating that the companies acted with malice in failing to warn customers about the dangers of MTBE contamination. As of fall 2006, hundreds of other lawsuits are still pending regarding MTBE contamination of public and private drinking water supplies. Most of these lawsuits have been consolidated into "multi-district litigation" before Judge Shira A. Scheindlin in the federal district court for the Southern District of New York.

The Energy Policy Act of 2005, passed in the House on April 21, 2005, did not include a provision for shielding MTBE manufacturers from water contamination lawsuits.

This provision was first proposed in 2003 and had been thought by some to be a priority of Tom DeLay and Rep. Joe Barton, chairman of the Energy and Commerce Committee. This bill also includes a provision that gives MTBE makers, including some major oil companies, $2 billion in transition assistance as MTBE is phased out over the next nine years.

Due to opposition in the Senate, the conference report dropped all MTBE provisions. The final bill was passed by both houses and signed into law by President Bush. The lack of MTBE liability protection is resulting in a switchover to the use of ethanol as a gasoline additive, which is in limited supply in April 2006. Some traders and consumer advocates are blaming this for an increase in gasoline prices.

Certain patents important in the manufacture of MTBE are not held by American companies; for example, United States patent 5536886, *Process for preparing alkyl ethers*, is owned by the Finnish company Neste. The same corporation also went on to patent the replacement of the MTBE process.

The United States Environmental Protection Agency (EPA) currently lists methyl tertiary butyl ether (MTBE) as a candidate for a maximum contaminant level (MCL) in drinking water. MCL's are determined by the EPA using toxicity data.

Methylsulphonylmethane

Methylsulphonylmethane (MSM, or dimethylsulphone) is an organic sulphur compound belonging to a class of chemicals known as sulphones. It occurs naturally in some primitive plants and is present in small amounts in many foods and beverages.

MSM is also known as dimethylsulphone, or $DMSO_2$, a name that reflects its close chemical relationship to dimethyl sulphoxide (DMSO), which differs only in the oxidation state of the sulphur atom. MSM is the primary metabolite of DMSO in humans, and it shares some of the properties of DMSO.

MSM is sold as a dietary supplement that is marketed with a variety of claims and is commonly used (often in combination with glucosamine and/or chondroitin) for helping to treat or prevent osteoarthritis.

Retail sales of MSM as a single ingredient in dietary supplements amounted to $115 million in 2003. However, clinical research on the medical use of the chemical in people is limited to a few pilot studies that have suggested beneficial effects.

Use as a Solvent

Because of its polarity and thermal stability, MSM is used industrially as a high-temperature solvent for both inorganic and organic substances. It is used as a medium for carrying out chemical reactions in polymerisation processes as well as to make pharmaceuticals, agrochemicals, paint, coating materials and biocides. According to US law enforcement officials, MSM is used to dilute methamphetamine in the illegal drug trade.

Effects on Health

The effects of supplemental methylsulphonylmethane in biology and medicine are poorly understood. Several researchers have suggested that MSM has anti-inflammatory effects (Morton *et al.* 1986; Childs, 1994; Murav'ev *et al.*, 1991). Any health effects of dimethyl sulphoxide (DMSO) may be mediated, at least in part, by MSM (Williams et al., 1966; Kocsis et al., 1975). Stanley W. Jacob, M.D., of the Oregon Health and Science University, claims to have used MSM to treat over 18,000 patients with a variety of ailments (Jacob & Appleton, 2003).

Clinical evidence for the usefulness of MSM is limited to animal studies and four published clinical studies in humans. These pilot studies of MSM have suggested some benefits, particularly for treatment of osteoarthritis. Further studies would be needed to test the usefulness of the chemical as a medical therapy.

Evidence from Clinical Trials

- *Osteoarthritis:* After several reports that MSM helped arthritis in animal models, a double-blind, placebo-controlled study suggested that 1500 mg per day MSM (alone or in combination with glucosamine sulphate) was helpful in relieving symptoms of knee osteoarthritis (Usha and Naidu 2004). Kim *et al.* then conducted a double-blind clinical trial of MSM for treatment of patients with osteoarthritis of the knee. Twenty-five patients took 6 g/day MSM and 25 patients took a placebo for 12 weeks. Ten patients did not complete the study, and intent-to-treat analysis was performed. Patients who took MSM had significantly reduced pain and improved physical functioning, without major adverse events (Kim *et al.*). No evidence of a more general anti-inflammatory effect was found, as there were no significant changes in two measures of systemic inflammation: C-reactive protein level and erythrocyte sedimentation rate. The authors cautioned that this short pilot study did not address the long-term safety and usefulness of MSM, but suggested that physicians should consider its use for certain osteoarthritis patients, and that long-term studies should be conducted (Kim *et al.* 2006).
- *Seasonal Allergic Rhinitis:* Barrager *et al.* evaluated the efficacy of MSM for hayfever (Barrager et al., 2002). Twenty-five subjects consumed 2,600 mg of MSM per day for 30 days, and a significant improvement in symptoms was observed compared to those taking a placebo. However, the study was not blinded. Also, no significant changes were observed in two indicators of inflammation (C-reactive protein and immunoglobulin E levels). The authors suggest that MSM is safe for short-term use and recommend that a larger, double-

blind study be performed to establish its usefulness in treating symptoms of seasonal allergic rhinitis.

- *Interstitial Cystitis:* In 1978, the FDA approved dimethylsulphoxide (DMSO) for instillation into the bladder as a treatment for interstitial cystitis. Since DMSO is metabolised to MSM by the body, it is possible that MSM is the active ingredient in DMSO treatments (Childs 1994).
- *Snoring:* Blum and Blum conducted a randomised, double-blind, placebo controlled clinical trial of an MSM-containing throat spray for snoring (Blum and Blum, 2004).

Pharmacology and toxicity

The LD50 (dose at which 50 per cent of test subjects are killed) of MSM is greater than 17.5 grams per kilogram of body weight. In rats, no adverse events were observed after daily doses of 2 g MSM per kg of body weight. In a 90-day follow-up study rats received daily MSM doses of 1.5 g/kg, and no changes were observed in terms of symptoms, blood chemistry, or gross pathology (Horvath *et al.*, 2002).

Nuclear magnetic resonance (NMR) studies have demonstrated that oral doses of MSM are absorbed into the blood and cross the blood-brain barrier (Rose *et al.*, 2000; Lin *et al.*, 2001). An NMR study has also found detectable levels of MSM normally present in the blood and cerebrospinal fluid, suggesting that it derives from dietary sources, intestinal bacterial metabolism, and the body's endogenous methanethiol metabolism (Engelke *et al.*, 2005).

The published clinical trials of MSM did not observe any serious side-effects of treatment, but there are no peer-reviewed data on the effects of long-term use in humans.

Calcination

Calcination (also referred to as Calcining) is thermal treatment process applied to ores and other solid materials in order to bring about a thermal decomposition, phase transition, or removal of a volatile fraction. The calcination process normally takes place at temperatures below the melting point of the product materials. Calcination is to be distinguished from roasting, in which more complex gas-solid reactions take place between the furnace atmosphere and the solids.

Industrial Processes

The process of calcination derives its name from its most common application, the decomposition of calcium carbonate (limestone) to calcium oxide (lime). The product of Calcination is usually referred to in general as "Calcine," regardless of the actual minerals undergoing thermal treatment. Calcination is carried out in furnaces or reactors (sometimes referred to as kilns) of various designs including shaft furnaces, rotary kilns, multiple hearth furnaces, and fluidised bed reactors.

Examples of calcination processes include the following:

- Decomposition of hydrated minerals, as in calcination of bauxite, to remove crystalline water as water vapour;
- Decomposition of carbonate minerals, as in the calcination of limestone to drive off carbon dioxide;
- Volatilisation of organic compounds (usually under inert or reducing atmospheres), as in petroleum coke calcination; and
- Heat treatment to effect phase transformations, as in conversion of anatase to rutile or devitrification of glass materials.

Calcination Reactions

Calcination reactions usually take place at or above the thermal decomposition temperature (for decomposition and volatilisation reactions) or the transition temperature (for phase transitions). This temperature is usually defined as the temperature at which the standard Gibb's free energy of reaction for a particular calcination reaction is equal to zero. For example, in limestone calcination, a decomposition process, the chemical reaction is $CaCO_3 = CaO + CO_2(g)$. The standard Gibb's free energy of reaction is approximated as $\Delta G°_r = 177{,}100 - 158\ T$ (J/mol). The standard free energy of reaction is zero in this case when the temperature, *T*, is equal to 848°C.

Examples of chemical decomposition reactions common in calcination processes, and their respective thermal decomposition temperatures include:

- $CaCO_3 \xrightarrow{848°C} CaO + CO_2$;

Alchemy

In alchemy, calcination was believed to be one of the 12 vital processes required for the transformation of a substance.

Alchemists distinguished two kinds of calcination, *actual* and *potential*. Actual calcination is that brought about by actual fire, from wood, coals, or other fuel, raised to a certain temperature. Potential calcination is that brought about by *potential* fire, such as corrosive chemicals; for example, gold was calcined in a reverberatory furnace with mercury and sal ammoniac; silver with common salt and alkali salt; copper with salt and sulphur; iron with sal ammoniac and vinegar; tin with antimony; lead with sulphur; and mercury with aqua fortis.

There was also *philosophical calcination*, which was said to occur when horns, hooves, etc., were hung over boiling water, or other liquor, till they had lost their mucilage, and were easily reducible into powder.

Geological

Calcination can also occur under layers of hot volcanic ash., e.g. Tungurahua volcano in Ecuador eruption in 2006.

Ceration

Ceration is a chemical process, a common practice in alchemy. It is one of the commonly accepted 12 vital alchemical processes.

Ceration is performed by continuously adding a liquid to

the substance while it is heated. This typically results in making the substance softer while giving it a waxy appearance.

Cohobation

In pre-modern chemistry and alchemy, cohobation was the process of repeated distillation of the same matter, with the liquid drawn from it; that liquid being poured again and again upon the matter left at the bottom of the vessel.

Cohobation is a kind of *circulation*, only differing from it in this, that the liquid is drawn off in cohobation, as in common distillation, and thrown back again; whereas in circulation, it rises and falls in the same vessel, without ever being drawn out.

Congelation

Congelation is the process by which something congeals, or thickens. This increase in viscosity can be achieved through a reduction in temperature or through chemical reactions. Sometimes the increase in viscosity is great enough to crystallise or solidify the substance in question. In alchemy, congelation is one of the 12 vital processes for transformation to occur.

Corporification

In pre-modern chemistry, corporification, or corporation, was the practice of recovering spirits into the same body, or at least into a body nearly the same, as that which they had before their spiritualisation.

Digestion (alchemy)

In alchemy, Digestion is a process in which gentle heat is applied to a substance over a period of several weeks.

This was traditionally performed by sealing a sample of the substance in a flask, and keeping the flask in fresh horse dung or sometimes in direct sunlight. Today, practitioners of alchemy use thermostat-controlled incubators.

Digestion is considered one of the 12 core alchemical processes and is "ruled", or "dominated", by the zodiacal sign of Leo.

Distillation

Distillation is a method of separating chemical substances based on differences in their volatilities. Distillation usually forms part of a larger chemical process, and is thus referred to as a unit operation.

Commercially, distillation has a number of uses. It is used to separate crude oil into more fractions for specific uses such as transport, power generation and heating. Water is distilled to remove impurities, such as salt from sea water. Air is distilled to its components such as oxygen for medical applications and helium for party balloons. The use of distillation on fermented solutions to produce distilled beverages with a higher alcohol content is perhaps the oldest form of distillation, known since ancient times.

Distillation was an invention of Greek alchemists in the first century AD and the later development of large-scale distillation apparatus occurred in response to demands for spirits. Hypathia of Alexandria is credited with having invented the distillation apparatus and the first exact description of apparatus for distillation is given by Zosimus of Alexandria, in the fourth century.

Distillation was developed further by Islamic alchemist Jabir ibn Hayyan in around 800 AD. He is credited with the invention of numerous chemical apparatus and processes that are still in use today. The design of the alembic has served as inspiration for some modern micro-scale distillation apparatus such as the Hickman stillhead.

As alchemy evolved into the science of chemistry, vessels called retorts became used for distillations. Both alembics and retorts are forms of glassware with long necks pointing to the side at a downward angle which acted as air-cooled condensers to condense the distillate and let it drip downward for collection.

Later, copper alembics were invented. Riveted joints were often kept tight by using various mixtures, for instance a dough made of rye flour. These alembics often featured a cooling system around the beak, using cold water for instance, which made the condensation of alcohol more efficient. These were called pot stills.

Today, the retorts and pot stills have been largely supplanted by more efficient distillation methods in most industrial processes. However, the pot still is still widely used for the elaboration of some fine alcohols, such as cognac, Scotch whisky and Ketel One Vodka. The unique shape of each pot still is said to give the alcohol a distinctive taste. Pot stills made of various materials (wood, clay, stainless steel) are also used by bootleggers in various countries. Small pot stills are also sold for the domestic production of flower water or essential oils.

Applications of Distillation

The application of distillation can roughly be divided in four groups: laboratory scale, industrial distillation, distillation of herbs for perfumery and medicinals (herbal distillate) and food processing. The latter two are distinct from the former two, in that in the distillation is not used as a true purification method, but more to transfer all volatiles from the source materials to the distillate.

The main difference between laboratory scale distillation and industrial distillation is that laboratory scale distillation is often performed batch-wise, whereas industrial distillation often occurs continuously. In batch distillation, the composition of the source material, the vapours of the distilling compounds and the distillate change during the distillation. In batch distillation, a still is charged (supplied) with a batch of feed mixture, which is then separated into its component fractions which are collected sequentially from most volatile to less volatile, with the bottoms (remaining least or non-volatile fraction) removed at the end. The still can then be recharged and the process repeated.

In continuous distillation, the source materials, vapours and distillate are kept at a constant composition by carefully replenishing the source material and removing fractions from both vapour and liquid in the system. This results in a better control of the separation process..

Idealised Distillation Model

It is a common misconception that in a solution, each component boils at its normal boiling point—the vapours of each component will collect separately and purely. This, however, does not occur even in an idealised system. Idealised models of distillation are essentially governed by Raoult's law and Dalton's law.

Raoult's law assumes that a component contributes to the total vapour pressure of the mixture in proportion to its percentage of the mixture and its vapour pressure when pure. If one component changes another component's vapour pressure, or if the volatility of a component is dependent on its percentage in the mixture, the law will fail.

Dalton's law states that the total vapour pressure is the sum of the vapour pressures of each individual component in the mixture. When a multi-component system is heated, the vapour pressure of each component will rise, thus causing the total vapour pressure to rise. When the total vapour pressure reaches the ambient pressure, boiling occurs and liquid turns to gas throughout the bulk of the solution. Note that a given mixture has one boiling point, when the components are mutually soluble.

The idealised model is accurate in the case of chemically similar liquids, such as benzene and toluene. In other cases, severe deviations from Raoult's law and Dalton's law are observed, most famously in the mixture of ethanol and water. These compounds, when heated together, form an azeotrope, in which the boiling temperature of the mixture is lower than the boiling temperature of each separate liquid. Virtually all liquids, when mixed and heated, will display azeotropic behaviour. Although there are computational methods that can be used to estimate the behaviour of a mixture of arbitrary components, the only way to obtain accurate vapour-liquid equilibrium data is by measurement.

It is not possible to completely purify a mixture of components by distillation, as this would require each component in the mixture to have a zero partial pressure. If ultra-pure products are the goal, then further chemical separation must be applied.

Batch Distillation

Heating an ideal mixture of two volatile substances A and B (with A having the higher volatility, or lower boiling point) in a batch distillation setup (such as in an apparatus depicted in the opening figure) until the mixture is boiling results in a vapour above the liquid which contains a mixture of A and B.

The ratio between A and B in the vapour will be different from the ratio in the liquid: the ratio in the liquid will be determined by how the original mixture was prepared, while the ratio in the vapour will be enriched in the more volatile compound, A. The vapour goes through the condenser and is removed from the system. This in turn means that the ratio of compounds in the remaining liquid is now different from the initial ratio (i.e. more enriched in B than the starting liquid).

The result is that the ratio in the liquid mixture is changing, becoming richer in component B. This causes the boiling point of the mixture to rise, which in turn results in a rise in the temperature in the vapour, which results in a changing ratio of A : B in the gas phase (as distillation continues, there is an increasing proportion of B in the gas phase). This results in a slowly changing ratio A : B in the distillate.

If the difference in vapour pressure between the two components A and B is large (generally expressed as the difference in boiling points), the mixture in the beginning of the distillation is highly enriched in component A, and when component A has distilled off, the boiling liquid is enriched in component B.

Continuous Distillation

In continuous distillation, the process is different from the above in that fractions are withdrawn from both the vapour and the liquid at such a speed that the combined ratio of the two fractions is exactly the same as the ratio in the starting mixture. In this way a stream of enriched component A and a stream of enriched component B is obtained. Moreover, a stream of crude mixture (which has the same ratio of A and B as the mixture in the still) can be added to the distilling mixture to replenish the liquid, meaning that the system can be run continuously.

General Improvements

Both batch and continuous distillations can be improved by making use of a fractionating column on top of the distillation flask. The column improves separation by providing a larger surface area for the vapour and condensate to come into contact. This helps it remain at equilibrium for as long as possible. The column can even exist of small subsystems ('trays' or 'dishes') which all contain an enriched, boiling liquid mixture, all with their own vapour-liquid equilibrium.

There are differences between laboratory-scale and industrial-scale fractionating columns, but the principles are the same. Examples of fractionating columns (in increasing efficacy) include:

- Air condenser
- Vigreux column (usually laboratory scale only)
- Packed column (packed with glass beads, metal pieces, or other chemically inert material)
- Spinning band distillation system

Laboratory Scale Distillation

Laboratory scale distillations are almost exclusively run as batch distillations. The device used in distillation, sometimes referred to as a still, consists at a minimum of a reboiler or pot in which the source material is heated, a condenser in which the heated vapour is cooled back to the liquid state, and a receiver in which the concentrated or purified liquid, called the distillate, is collected. Several laboratory scale techniques for distillation exist.

Simple Distillation

In simple distillation, all the hot vapours produced are immediately channelled into a condenser which cools and condenses the vapours. Thus, the distillate will not be pure - its composition will be identical to the composition of the vapours at the given temperature and pressure, and can be computed from Raoult's law.

As a result, simple distillation is usually used only to separate liquids whose boiling points differ greatly (rule of thumb is 25°C), or to separate liquids from involatile solids or oils. For these cases, the vapour pressures of the components are usually sufficiently different that Raoult's law may be neglected due to the insignificant contribution of the less volatile component. In this case, the distillate may be sufficiently pure for its intended purpose.

Fractional Distillation

For many cases, the boiling points of the components in the mixture will be sufficiently close that Raoult's law must be taken into consideration. Thus, fractional distillation must be used in order to separate the components well by repeated vaporisation-condensation cycles within a packed fractionating column.

As the solution to be purified is heated, its vapours rise to the fractionating column. As it rises, it cools, condensing on the condenser walls and the surfaces of the packing material. Here, the condensate continues to be heated by the rising hot vapours; it vaporises once more. However, the composition of the fresh vapours are determined once again by Raoult's law. Each vaporisation-condensation cycle (called a *theoretical plate*) will yield a purer solution of the more volatile component. In reality, each cycle at a given temperature does not occur at exactly the same position in the fractionating column; *theoretical plate* is thus a concept rather than an accurate description.

More theoretical plates lead to better separations. A spinning band distillation system uses a spinning band of Teflon or metal to force the rising vapours into close contact with the descending condensate, increasing the number of theoretical plates.

Steam Distillation

Like vacuum distillation, steam distillation is a method for distilling compounds which are heat-sensitive. This process involves using bubbling steam through a heated mixture of the raw material. By Raoult's law, some of the target compound will vaporise (in accordance with its partial pressure). The vapour mixture is cooled and condensed, usually yielding a layer of oil and a layer of water.

Steam distillation of various aromatic herbs and flowers can result in two products; an essential oil as well as a watery herbal distillate. The essential oils are often used in perfumery and aromatherapy while the watery distillates have many applications in aromatherapy, food processing and skin care.

Fermentation

Fermentation is a process of energy production in a cell in an anaerobic environment (with no oxygen present). In common usage fermentation is a type of anaerobic respiration, however a more strict definition exists which defines fermentation as respiration in an anaerobic environment with no external electron acceptor.

Sugars are the common substrate of fermentation, and typical examples of fermentation products are ethanol, lactic acid, and hydrogen. However, more exotic compounds can be produced by fermentation, such as butyric acid and acetone. Yeast famously carries out fermentation in the production of ethanol in beers, wines and other alcoholic drinks, along with the production of large quantities of carbon dioxide. Anaerobic respiration in mammalian muscle under periods of intense exercise (which has no external electron acceptor) is, under the strict definition, a type of fermentation.

French chemist Louis Pasteur was the first zymologist, when in 1857 he connected yeast to fermentation. Pasteur originally defined fermentation as *respiration without air*.

The German Eduard Buchner, winner of the 1907 Nobel Prize in chemistry, later determined that fermentation was actually caused by a yeast secretion that he termed *zymase*.

The research efforts undertaken by the Danish Carlsberg scientists greatly accelerated the gain of knowledge about yeast and brewing. The Carlsberg scientists are generally acknowledged with jump-starting the entire field of molecular biology.

Reaction

The reaction of fermentation differs according to the sugar being used and the product produced. Below the sugar will be glucose ($C_6H_{12}O_6$) the simplest sugar, and the product will be ethanol ($2C_2H_5OH$). This is one of the fermentation reactions carried out by yeast, and is used in food production.

Chemical Equation

$$C_6H_{12}O_6 \longrightarrow 2C_2H_5OH + 2CO_2 + 2\ ATP$$

(Energy Released:118 kJ/mol)

The actual biochemical pathway the reaction takes varies depending on the sugars involved, but commonly involves part of the glycolysis pathway, which is shared with the early stages of aerobic respiration in most organisms. The later stages of the pathway vary considerably depending on the final product.

Energy Source in Anaerobic Conditions

Fermentation is thought to have been the primary means of energy production in earlier organisms before oxygen was at high concentration in the atmosphere and thus would represent a more ancient form of energy production in cells.

Fermentation products contain chemical energy (they are not fully oxidised) but are considered waste products since they cannot be metabolised further without the use of oxygen (or other more highly-oxidised electron acceptors). A consequence is that the production of ATP by fermentation is less efficient than oxidative phosphorylation, where pyruvate is fully oxidised to carbon dioxide. Fermentation produces two ATP molecules per molecule of glucose compared to approximately 36 by aerobic respiration.

Aerobic glycolysis is a method employed by muscle cells for the production of lower-intensity energy over a longer period of time when oxygen is plentiful. Under low-oxygen conditions, however, vertebrates use the less-efficient but faster *anaerobic glycolysis* to produce ATP. The speed at which ATP is produced is about 100 times that of oxidative phosphorylation. While fermentation is helpful during short, intense periods of exertion, it is not sustained over extended periods in complex aerobic organisms. In humans, for example, lactic acid fermentation provides energy for a period ranging from 30 seconds to 2 minutes.

The final step of fermentation, the conversion of pyruvate to fermentation end-products, does not produce energy. However, it is critical for an anaerobic cell since it regenerates nicotinamide adenine dinucleotide (NAD^+), which is required for glycolysis. This is important for normal cellular function, as glycolysis is the only source of ATP in anaerobic conditions.

Products

Products produced by fermentation are actually waste products produced during the reduction of pyruvate to regenerate NAD^+ in the absence of oxygen. Bacteria generally produce acids. Vinegar (acetic acid) is the direct result of bacterial metabolism (Bacteria need oxygen to convert the alcohol to acetic acid). In milk, the acid coagulates the casein, producing curds. In pickling, the acid preserves the food from pathogenic and putrefactive bacteria.

When yeast ferments, it breaks down the glucose ($C_6H_{12}O_6$) into exactly two molecules of ethanol (C_2H_6O) and two molecules of carbon dioxide (CO_2).

- Ethanol fermentation (performed by yeast and some types of bacteria) breaks the pyruvate down into ethanol and carbon dioxide. It is important in bread-making, brewing, and wine-making. When the ferment has a high concentration of pectin, minute quantities of methanol can be produced. Usually only one of the products is desired; in bread the alcohol is baked out, and in alcohol production the carbon dioxide is released into the atmosphere.
- Lactic acid fermentation breaks down the pyruvate into lactic acid. It occurs in the muscles of animals when they need energy faster than the blood can supply oxygen. It also occurs in some bacteria and some fungi. It is this type of bacteria that converts lactose into lactic acid in yogurt, giving it its sour taste.

In vertebrates, during intense exercise, cellular respiration will deplete oxygen in the muscles faster than it can be replenished. An associated burning sensation in muscles has been attributed lactic acid causing a decrease in the pH during a shift to anaerobic glycolysis.

While this does partially explain acute muscle soreness, lactic acid may also help delay muscle fatigue, although, eventually the lower pH will inhibit enzymes involved in glycolysis. Contrary to currently popular belief, the lactic acid is not the primary causes for the drop in pH, but rather ATP-derived hydrogen ions.

Delayed onset muscle soreness cannot be attributed to the lactic acid and other waste products as they are quickly removed after exercise. It is actually due to microtrauma of the muscle fibres. Eventually the liver metabolises the lactic acid back to pyruvate.

Hydrogen gas is produced in many types of fermentation

(mixed acid fermentation, butyric acid fermentation, caproate fermentation, butyric acid fermentation, butanol fermentation, glyoxylate fermentation), as a way to regenerate NAD^+ and FAD from NADH and $FADH_2$. Electrons are transferred to ferredoxin, which in turn is oxidised by hydrogenase, producing H_2. Hydrogen gas is a substrate for methanogens and sulphate reducers, who keep the concentration of hydrogen sufficiently low to allow the production of such an energy-rich compound.

Some anaerobic eukaryotic microorganisms also produce hydrogen gas, in their hydrogenosomes. The concentration of hydrogen gas is kept low by symbionts such as methanogens that reside in the cytosol of the eukaryot.

Enzymology

Enzymology is the scientific term for yeast oriented fermentation. It deals with the biochemical processes involved in fermentation, with yeast selection and physiology, and with the practical issues of brewing. Enzymology is occasionally known as *zymology* or *zymurgy*.

Fermentation (food)

Fermentation typically refers to the conversion of sugar to alcohol using yeast under anaerobic conditions. A more general definition of fermentation is the chemical conversion of carbohydrates into alcohols or acids. When fermentation stops prior to complete conversion of sugar to alcohol, a stuck fermentation is said to have occurred. The science of fermentation is known as zymology.

Fermentation usually implies that the action of the microorganisms is desirable, and the process is used to produce wine, beer, hard cider, and vinegar. Fermentation is also employed in preservation to create lactic acid in sour foods such as pickled cucumbers, kimchi and yogurt. Occasionally wines are enhanced through the process of cofermentation.

Since fruits ferment naturally, fermentation precedes human history. Since pre-historic times, however, humans have been taking control of the fermentation process. The earliest evidence of wine-making dates from 6000 BC, in Georgia (the former Soviet Republic). 7000-year old jars of wine have been excavated in the Zagros Mountains in Iran, which are now on display at the University of Pennsylvania. There is strong evidence that people were fermenting beverages in Babylon circa 5000 BC, ancient Egypt circa 3150 BC, pre-Hispanic Mexico circa 2000 BC, and Sudan circa 1500 BC. There is also evidence of leavened bread in ancient Egypt circa 1500 BC and of milk fermentation in Babylon circa 3000 BC. The Chinese were probably the first to develop vegetable fermentation.

Uses

The primary benefit of fermentation is the conversion of sugars and other carbohydrates, e.g., converting juice into wine, grains into beer, carbohydrates into carbon dioxide to leaven bread, and sugars in vegetables into preservative organic acids.

Food fermentation has been said to serve five main purposes:

1. Enrichment of the diet through development of a diversity of flavours, aromas, and textures in food substrates.
2. Preservation of substantial amounts of food through lactic acid, alcohol, acetic acid and alkaline fermentations.
3. Biological enrichment of food substrates with protein, essential amino acids, essential fatty acids, and vitamins.
4. Detoxification during food-fermentation processing.
5. A decrease in cooking times and fuel requirements.

Fermentation has some uses exclusive to foods. Fermentation can produce important nutrients or eliminate anti-nutrients. Food can be preserved by fermentation, since fermentation uses up food energy and can make conditions unsuitable for undesirable microorganisms.

For example, in pickling the acid produced by the dominant bacteria inhibit the growth of all other microorganisms. Depending on the type of fermentation, some products (e.g., fusel alcohol) can be harmful to people's health.

In alchemy, fermentation is often the same as putrefaction, meaning to allow the substance to naturally rot or decompose.

Stuck Fermentation

A *stuck fermentation* is where a fermentation has stopped before completion; i.e., before the anticipated percentage of sugars has been converted by yeast into alcohol or carbohydrates into carbon dioxide.

Typically, a stuck fermentation may be caused by: (1) insufficient or incomplete nutrients required to allow the yeast to complete fermentation; (2) low temperatures, or temperature changes which have caused the yeast to stop working early; or (3) a percentage of alcohol which has grown too high for the particular yeast chosen for the fermentation.

Corrections to stuck fermentations may include: (1) repitching a different yeast; (2) incorporation of nutrients in conjunction with the repitched yeast; (3) restoration of accommodative temperatures for the given yeast.

Putrefaction

Putrefaction is the decomposition of animal proteins, especially by anaerobic microorganisms, described as putrefying bacteria.

Decomposition is a more general process. Putrefaction usually results in amines such as putrescine and cadaverine, which have a putrid odour. Material that is subject to putrefaction is called putrescible.

In alchemy, putrefaction is the same as fermentation, basically meaning to allow the substance to rot or decompose, sometimes with a small sample of the desired original pure material to act as a "seed".

Coatings

Conversion Coating: Conversion coatings are coatings for metals where the part surface is converted into the coating with a chemical or electrochemical process. Examples include chromate conversion coatings, phosphate conversion coatings, bluing, oxide coatings on steel, and anodising. They are used for corrosion protection, increased surface hardness, to add decorative colour and as paint primers. Conversion coatings may be very thin, on the order of 0.00001. Thick coatings, up to 0.002, are usually built up on aluminium alloys, either by anodising or chromate conversion.

Chromate Conversion Coating

Chromate conversion coatings are a type of conversion coating applied to passivate aluminium, zinc, cadmium, copper, silver, magnesium, tin and their alloys to slow corrosion. The process uses various toxic chromium compounds which may include hexavalent chromium. The industry is developing less toxic alternatives in order to comply with substance restriction legislation. One alternative is trivalant chromate conversion which is not as effective but less environmentally damaging.

Chromating is commonly used on zinc-plated parts to protect the zinc from white corrosion, which is primarily a cosmetic issue. It cannot be applied directly to steel or iron, and does not enhance zinc's anodic protection of the underlying steel from brown corrosion.

It is also commonly used on aluminium alloy parts in the aircraft industry where it is often called *chemical film*. It has additional value as a primer for subsequent organic coatings, as untreated metal, especially aluminium, is difficult to paint or glue. Chromated parts retain their electrical conductivity to varying degrees, depending on coating thickness. The process may be used to add colour for decorative or identification purposes.

Chromate coatings are soft and gelatinous when first applied but harden and become hydrophobic as they age.

Curing can be accelerated by heating up to 70°C, but higher temperatures will gradually damage the coating over time. Some chromate conversion processes use brief degassing treatments at temperatures of up to 200°C.

Coating thickness from a few nanometres to several hundred nanometres can be produced, but the Alodine and Modified Bauer-Vogel coatings on aluminium are typically a few micrometers thick.

The protective effect of chromate coatings on zinc is indicated by colour, progressing from clear/blue to yellow, gold, olive drab and black. Darker coatings generally provide more corrosion resistance. Chromate conversion coatings are common on everyday items such as hardware and tools and usually have a distinctive yellow colour. Steel parts must be plated with zinc or cadmium prior to chromating.

Phosphate coatings on iron and steel may also be treated with a chromic acid rinse to enhance the phosphate coating.

Standard Specifications

Mil-C-5541 specifies chromate conversion of aluminium alloy parts.

ASTM B633 specify zinc plating plus chromate conversion on iron and steel parts.

Composition

The composition of chromate conversion solutions varies widely depending on the material to be coated and the desired effect. Most solution compositions are proprietary.

The widely used Cronak process for zinc and cadmium consists of 5-10 seconds of immersion at room temperature in a solution of 182 g/l sodium dichromate crystals ($Na_2Cr_2O_72H_2O$) and 6 ml/l concentrated sulphuric acid.

Iridite 14-2, a chromate conversion coating for aluminium, contains chromium oxide barium nitrate and sodium silico fluoride.

Plasma Electrolytic Oxidation

Plasma electrolytic oxidation (PEO), or microarc oxidation (MAO), is an electrochemical surface treatment process for generating oxide coatings on metals. It is similar to anodising, but it employs higher potentials, so that discharges occur. This process can be used to grow thick (tens or hundreds of micrometers), largely crystalline, oxide coatings on metals such as aluminium, magnesium and titanium. Because they can present high hardness and a continuous barrier, these coatings can offer protection against wear, corrosion or heat as well as electrical insulation.

The coating is a chemical conversion of the substrate metal into its oxide, and grows both inwards and outwards from the original metal surface. Because it is a conversion coating, rather than a deposited coating (such as a coating formed by plasma spraying), it has excellent adhesion to the substrate metal. A wide range of substrate alloys can be coated, including all wrought aluminium alloys and most cast alloys, although high levels of silicon can reduce coating quality.

Process

Metals such as aluminium naturally form a passivating oxide layer which provides moderate protection against corrosion. The layer is strongly adherent to the metal surface, and it will regrow quickly if scratched off. In conventional anodising, this layer of oxide is grown on the surface of the metal by the application of electrical potential, while the part is immersed in an acidic electrolyte.

In plasma electrolytic oxidation, higher potentials are applied. For example, in the plasma electrolytic oxidation of aluminium, at least 200 V must be applied. This exceeds the dielectric breakdown potential of the growing oxide film, and discharges occur. These discharges sinter and densify the growing oxide and also partially convert it from amorphous alumina into crystalline forms such as corundum (α -Al_2O_3). As a result, mechanical properties such as hardness and toughness are enhanced.

Equipment used

The part to be coated is immersed in a bath of electrolyte which usually consists of a dilute alkaline solution such as KOH. It is electrically connected, so as to become one of the electrodes in the electrochemical cell, with the other, being a stainless steel counter-electrode, often the wall of the bath itself. Potentials of over 200V are applied between these two electrodes. These may be continuous or pulsed DC (in which case the part is simply an anode in Direct current operation), or alternating pulses (alternating current operation).

Coating Properties

Plasma electrolytic oxide coatings are generally recognised for high hardness, wear resistance, and corrosion resistance. However, the coating properties are highly dependent on the substrate used, as well as on the composition of the electrolyte and the electrical regime used. Even on aluminium, the coating properties can vary strongly according to the exact alloy composition. Extensive work is being pursued by Prof. T. W. Clyne at the University of Cambridge to investigate microstructural and mechanical characteristics of PEO coatings.

Phosphate Conversion Coating

Phosphate coatings are used on steel parts for corrosion resistance, lubricity, or as a foundation for subsequent coatings or painting. It serves as a conversion coating in which a dilute solution of phosphoric acid, which is applied via spraying or immersion, chemically reacts with the surface of the part being coated to form a layer of insoluble, crystalline phosphates. Phosphate conversion coatings can also be used on aluminium, zinc, cadmium, silver and tin.

Phosphate and Oil Coatings

Phosphate coatings are porous, so they must be treated with oils or other sealers in order to provide corrosion resistance. P&O (phosphate and oil) coatings are frequently used for this purpose, for lubricity and to prevent galling.

Base for Painting and Coating

Most phosphate coatings serve as a surface preparation for further coating and/or painting, a function it performs effectively with excellent adhesion and electric isolation. The porosity allows the additional materials to seep into the phosphate coating and become mechanically interlocked after drying. The dielectric nature will electrically isolate anodic and cathodic areas on the surface of the part, minimising underfilm corrosion that sometimes occurs at the interface of the paint/coating and the substrate.

Types

The main types of phosphate coatings are manganese, iron and zinc. Manganese phosphates are used both for corrosion resistance and lubricity and are applied only by immersion. Iron phosphates are typically used as a base for further coatings or painting and are applied by immersion or by spraying. Zinc phosphates are used for rust proofing, lubricity (P&O), and as a paint/coating base and can also be applied by immersion or spraying.

Performance

The performance of the phosphate coating is significantly dependent on the crystal structure as well as the weight. For example, a microcrystalline structure is usually optimal for corrosion resistance or subsequent painting. A coarse grain structure impregnated with oil, however, may be the most desirable for wear resistance. These factors are controlled by selecting the appropriate phosphate solution, using various additives, and controlling bath temperature, concentration, and phosphating time.

Chemistry

Phosphating solutions are based on phosphoric acid, which attacks the surface of the substrate and caused some of the metal atoms to go into solution. This neutralises the acid solution in a thin zone along the metal-bath interface, resulting in a lower solubility of the metal phosphates and the subsequent precipitation of them onto the surface of the workpiece by electrostatic forces.

In addition to phosphoric acid, phosphating solutions contain divalent metal phosphates and chemicals that serve to increase the rate of coating buildup by removing hydrogen from the surface of the substrate. Hydrogen is a by-product of the chemical reaction and can stall the process by aggregating to preventing the bath from coming into contact with the metal.

Process

A typical phosphating procedure consists of performing the following on the substrate:

1. Cleaning the surface
2. Rinsing
3. Surface activation (in some cases)
4. Phosphating
5. Rinsing
6. Neutralising rinse (optional)
7. Drying
8. Application of supplemental sealers, oil, etc.

ASTM B633 Type IV specifies zinc plating plus phosphate conversion on iron and steel parts.

Paint

Paint is any liquid, liquifiable, or mastic composition which after application to a substrate in a thin layer is converted to an opaque solid film.

Paint is used to protect, decorate (such as adding colour), or add functionality to an object or surface by covering it with a pigmented coating. An example of protection is to retard corrosion of metal. An example of decoration is to add festive trim to a room interior. An example of added functionality is to modify light reflection or heat radiation of a surface.

As a verb, painting is the application of paint. Someone who paints artistically is usually called a painter, while someone who paints commercially is often referred to as a painter and decorator, or house painter.

Paint can be applied to almost any kind of object. It is used, among many other uses, in the production of art, in industrial coating, as a driving aid (road surface marking), or as a barrier to prevent corrosion or water damage. Paint is a semi-finished product, as the final product is the painted article itself.

Components

There are three primary components to a paint:

- Pigments
- Binder, also known as non-volatile vehicle or resin
- Vehicle, also known as volatile vehicle, also called solvent

Pigment

Pigments impart such qualities as colour and opacity (sometimes inappropriately called 'hiding'), and influence properties such as gloss, film flow, and protective abilities. Pigment can generally be categorised into two main types: Prime, or hiding, pigments, and Inert, or extender, pigments.

The main modern white hiding pigment is Titanium dioxide. Zinc oxide is a weaker white pigment with some important usages. Colour hiding pigments fall also into two main categories, those being Inorganic, mostly duller earth tone colours, and Organic, generally brighter but more expensive colours.

Inert pigments break down into natural or synthetic types. Natural pigments include various clays, calcium carbonate, mica, silicas, and talcs. Synthetics would include calcined clays, blanc fix, precipitated calcium carbonate, and synthetic silicas.

Hiding pigments, in making paint opaque, also protect the substrate from the harmful effects of ultraviolet light.

Some pigments are toxic, such as the lead pigments that were used in lead paint. Paint manufacturers began replacing white lead pigments with the less toxic substitute, which can even be used to colour food, titanium white (titanium dioxide), even before lead was functionally banned in paint for residential use in 1978 by the US Consumer Product Safety Commission.

Titanium dioxide was first used in paints in the 19th century. The titanium dioxide used in most paints today is often coated with silicon or aluminium oxides for various reasons such as better exterior durability, or better hiding performance (opacity) via better efficiency promoted by more optimal spacing within the paint film. Opacity is also improved by optimal sizing of the titanium dioxide particles.

Binder

The binder, or resin, is the actual film forming component of paint. It imparts adhesion, binds the pigments together, and strongly influences such properties as gloss potential, exterior durability, flexibility, and toughness.

Binders include synthetic or natural resins such as acrylics, polyurethanes, polyesters, melamine resins, epoxy, or oils.

Binders can be categorised according to drying, or curing, mechanism. The four most common are simple solvent evaporation, oxidative cross-linking, catalysed polymerisation, and coalescence.

Note that drying and curing are two different processes. Drying generally refers to evaporation of vehicle, whereas curing refers to polymerisation of the binder. Depending on chemistry and composition, any particular paint may undergo either, or both processes. Thus, there are paints that dry only, those that dry then cure, and those that do not depend on drying for curing.

Paints that dry by simple solvent evaporation contain a solid binder dissolved in a solvent; this forms a solid film when the solvent evaporates, and the film can redissolve in the solvent again. Classic nitrocellulose lacquers fall into this category, as do non-grain raising stains composed of dyes dissolved in solvent.

Paints that cure by oxidative cross-linking are generally single package coatings that when applied, the exposure to oxygen in the air starts a process that cross-links and polymerises the binder component.

Paints that cure by catalysed polymerisation are generally two package coatings that polymerise by way of a chemical reaction initiated by mixing resin and hardener, and which cure by forming a hard plastic structure. Depending on composition they may need to dry first, by evaporation of solvent. Classic two package epoxies or polyurethanes would fall into this category.

Latex paints cure by a process called coalescence where first the water, and then the trace, or coalescing, solvent, evaporate and draw together and soften the latex binder particles together and fuse them together into irreversibly bound networked structures, so that the paint will not redissolve in the solvent/water that originally carried it.

Recent environmental requirements restrict the use of Volatile Organic Compounds (VOCs), and alternative means of curing have been developed, particularly for industrial purposes. In UV curing paints, the solvent is evaporated first, and hardening is then initiated by ultraviolet light.

Vehicle or Solvent

The main purpose of the vehicle is to adjust the viscosity of the paint. It is volatile and does not become part of the paint film. It can also control flow and application properties. It's main function is as the carrier for the non-volatile components.

Water is the main vehicle for water based paints.

Solvent based, sometimes called oil based, paints can have various combinations of solvents as the vehicle, including aliphatics, aromatics, alcohols, and ketones. These include organic solvents such as petroleum distillate, alcohols, ketones, esters, glycol ethers, and the like. Sometimes volatile low-molecular weight synthetic resins also serve as diluents.

Additives

Besides the three main categories of ingredients, paint can have a wide variety of miscellaneous additives, usually added in very small amounts. Some examples include additives to improve wet edge, to impart antifreeze properties, control foaming, control skinning, fight bacterial growth, or improve pigment stability.

Application

Paint can be applied as a solid, a gaseous suspension (aerosol) or a liquid. Techniques vary depending on the practical or artistic results desired. As a solid (usually used in industrial and automotive applications), the paint is applied as a very fine powder, then baked at high temperature. This melts the powder and causes it to adhere (stick) to the surface. The reasons for doing this involve the chemistries of the paint, the surface itself, and perhaps even the chemistry of the substrate (the overall object being painted).

As a gas or as a gaseous suspension, the paint is suspended in solid or liquid form in a gas that is sprayed on an object. The paint sticks to the object. The reasons for doing this include:

- The application mechanism is air and thus no solid object ever touches the object being painted;
- The distribution of the paint is very uniform so there are no sharp lines;
- It is possible to deliver very small amounts of paint or to paint very slowly;
- A chemical (typically a solvent) can be sprayed along with the paint to dissolve together both the delivered paint and the chemicals on the surface of the object being painted; and
- Some chemical reactions in paint involve the orientation of the paint molecules.

In the liquid application, paint can be applied by direct application using brushes, paint rollers, blades, other instruments, or body parts. Examples of body parts include fingerpainting, where the paint is applied by hand, whole-body painting (popular in the 1960s avant-garde movement), and cave painting, in which a pigment (usually finely-ground charcoal) is held in the mouth and spat at a wall (Note: some paints are toxic and might cause death or permanent injury).

Rollers generally have a handle that allows for different lengths of poles which can be attached to allow for painting at different heights. Generally, roller application takes two coats for even colour. A roller with a thicker nap is used to apply paint on uneven surfaces. Edges are often finished with an angled brush.

After liquid paint is applied, there is an interval during which it can be blended with additional painted regions (at the "wet edge") called "open time." The open time of an oil or alkyd-based emulsion paint can be extended by adding white spirit, similar glycols such as Dowanol (propylene glycol ether) or commercial open time prolongers. This can also facilitate the mixing of different wet paint layers for aesthetic effect. Latex and acrylic emulsions require the use of drying retardants suitable for water-based coatings.

Paint may also be applied by flipping the paint, dripping, or by dipping an object in paint.

Interior/exterior house paint tends to separate when stored, the heavier components settling to the bottom. It should be mixed before use, with a flat wooden stick or a paint mixing accessory; pouring it back and forth between two containers is also an effective manual mixing method. Paint stores have machines for mixing the paint by shaking it vigorously in the can for a few minutes.

Water-based paints tend to be the safest, and easiest to clean up after using — the brushes and rollers can be cleaned with soap and water.

It is difficult to reseal the paint container and store the paint well for a long period of time. Store upside down, for a good seal, in a cool dry place. Protect from freezing.

Proper disposal of paint is a challenge. Avoid acquiring excess paint. Look for suitable recycled paint before buying more. Try to find recycled uses for your left over paint. Paints of similar chemistry can be mixed to make a larger amount of a uniform colour. Old paint may be usable for a primer coat or an intermediate coat.

If you must dispose of paint, small quantities of water based paint can be carefully dried by leaving the lid off until it solidifies, and then disposing with normal trash. But oil based paint should be treated as hazardous waste, and disposed of according to local regulations.

Product Variants

- Primer is a preparatory coating put on materials before painting. Priming ensures better adhesion of paint to the surface, increases paint durability, and provides additional protection for the material being painted.
- Varnish and shellac provide a protective coating without changing the colour. They are paints without pigment.
- Wood stain is a type of paint that is very "thin," that is, low in viscosity, and formulated so that the pigment penetrates the surface rather than remaining in a film on top of the surface. Stain is predominantly pigment or dye and solvent with little binder, designed primarily to add colour without providing a surface coating.
- Lacquer is usually a fast-drying solvent-based paint or varnish that produces an especially hard, durable finish.
- An enamel paint is a paint that dries to an especially hard, usually glossy, finish. Enamel can be made by adding varnish to oil-based paint.
- A roof coating is a fluid applied membrane which has elastic properties that allows it to stretch and return to

their original shape without damage. It provides UV protection to polyurethane foam and is widely used as part of a roof restoration system.

- Fingerpaint.
- Inks are similar to paints, except they are typically made using dyes exclusively (no pigments), and are designed so as not to leave a thick film of binder.
- Titanium dioxide is extensively used for both house paint and artist's paint, because it is permanent and has good covering power. Titanium oxide pigment accounts for the largest use of the element. Titanium paint is an excellent reflector of infrared, and is extensively used in solar observatories where heat causes poor seeing conditions.
- Anti-Graffiti paints are used to defeat the marking of surfaces by graffiti artists. There are two categories, sacrificial and non-bonding. Sacrificial coatings are clear coatings that allow the removal of graffiti, usually by pressure washing the surface with high-pressure water, removing the graffiti, and the coating (hence, sacrificed.) They must be reapplied afterwards for continued protection. This is most commonly used on natural-looking masonry surfaces, such as statuary and marble walls, and on rougher surfaces that are difficult to clean. Non-bonding coatings are clear, high-performance coatings, usually catalysed polyurethanes, that allow the graffiti very little to bond to. After the graffiti is discovered, it can be removed with the use of a solvent wash, without damaging the underlying substrate or protective coating. These work best when used on smoother surfaces, and especially over other painted surfaces, including murals.
- Anti-climb paint is a non-drying paint that appears normal while still being extremely slippery. It is usually used on drainpipes and ledges to deter burglars and vandals from climbing them, and is found in many

public places. When a person attempts to climb objects coated with the paint, it rubs off onto the climber, as well as making it hard for them to climb.

- No-VOC paints, which are solvent-free paints that do not contain volatile organic compounds, have been available since the late 1980s. Low VOC paints, which typically contain anywhere between 0.3 - 5.0 per cent VOCs as coalescent, or coalescing solvent have been available since the 1960s.

History

Cave paintings drawn with red and yellow ochre, haematite, manganese oxide and charcoal may have been made by early homo sapiens as long as 40,000 years ago.

Ancient painted walls, to be seen at Dendera, Egypt, although exposed for many ages to the open air, still possess a perfect brilliancy of colour, as vivid as when painted, perhaps 2,000 years ago. The Egyptians mixed their colours with some gummy substance, and applied them detached from each other without any blending or mixture. They appeared to have used six colours: white, black, blue, red, yellow, and green. They first covered the field entirely with white, upon which they traced the design in black, leaving out the lights of the ground colour. They used minimum for red, and generally of a dark tinge.

Pliny mentions some painted ceilings in his day in the town of Ardea, which had been executed at a date prior to the foundation of Rome. He expresses great surprise and admiration at their freshness, after the lapse of so many centuries.

Paint was made with the yolk of eggs and therefore, the substance would harden and stick onto the surface applied.

Industrial Coating

An industrial coating is a paint or coating defined by its protective, rather than its aesthetic properties, although it can provide both.

The most common use of industrial coatings is for corrosion control of steel or concrete. Other functions include intumescent coatings for fire resistance. The most common polymers used in industrial coatings are polyurethane and epoxy. Another highly common polymer used in industrial coating is a fluoropolymer. There are many types of industrial coatings including inorganic zinc, phosphate, and Xylan.

NACE International and The Society for Protective Coatings (SSPC) are professional organisations involved in the industrial coatings industry.

Silicate Mineral Paint

Mineral Paints are mineral based coatings formulated with potassium silicate or sodium silicate, otherwise known as waterglass as the binder, combined with inorganic, alkaline resistant pigments. They are fully inorganic (containing no organic solvents) and are non-offgassing. Mineral paints petrify, by binding to any silicates within the substrate, forming a microcrystalline structure and a breathable finish. They are more of a stain, that becomes integral to the substrate, rather than a coating. They are alkaline and therefore inhibit microbiotic growth, and reduce carbonisation of cementitious materials.

The majority of non-toxic concrete stains and limestone restoration products are waterglass based. Mineral paints are also used as a non-toxic wood preservative.

The difference between the use of sodium silicate and potassium silicate as a binder is mainly geographic. The western hemisphere mainly produces sodium silicate, while Europe produces potassium silicate

Possible substrates:

- Masonry
- Stone
- Concrete
- Lime plasters
- Earthen plasters

Applications:

- Environmentally friendly, non-toxic applications
- High durability, especially on masonry products, and lightfast
- Breathable finish
- Acid rain resistance
- Antifungal properties
- Reduces carbonisation of cement based materials

Fusion Bonded Epoxy Coating

Fusion bonded epoxy coating, also known as fusion-bond epoxy powder coating and commonly referred to as FBE coating, is an epoxy based powder coating that is widely used to protect various sizes of steel pipes used in pipeline construction, concrete reinforcing rebars and on a wide variety of piping connections, valves, etc. from deterioration due to corrosion. FBE coatings are thermoset polymer coatings. They come under the category of 'protective coatings' in paints and coating nomenclature.

The name 'fusion-bond epoxy' is derived from the way of resin cross-linking and their method of application which is different from that of a conventional liquid paint. FBE coatings are in the form of dry powder at normal atmospheric temperatures. The resin and hardener parts in the dry powder remain unreacted at normal storage conditions. At typical coating application temperatures, usually in the range of 180 to 250°C, the contents of the powder melt and transform to a liquid form. The liquid FBE film "wets and flows on to" the steel surface, on which it is applied, and soon becomes a solid coating by chemical cross-linking, assisted by heat. This process is known as "fusion bonding". The chemical cross-linking reaction taking place in this case is "irreversible", which means once the curing takes place, the coating cannot be converted back into its original form by any means. Application of further heating will not "melt" the coating and thus it is known as a "thermoset" coating. World's leading FBE manufacturers are Jotun Powder Coatings, 3M, DuPont, Akzo Nobel, BASF and Rohm & Haas.

Since their introduction as a protective coating in early 1960s, FBE coating formulations had gone through vast improvements and developments. Today, various types of FBE coatings, which are tailor made to meet the varying requirements of the industry are available in the market.

Following recent research works by various FBE manufacturers, FBEs are available as stand-alone coatings as well as a part in multi-layers. FBE coatings with different chemical and physical properties are available to suit coating application on the main body of the pipes, internal surface, girth welds as well as on fittings. Similarly, variations are also available to match different ranges of pipeline service conditions.

Chemistry of FBE Coatings

Essential components of a powder coating are

1. Resin
2. Hardener or curing agent
3. Fillers and extenders
4. Colour pigments

Since their introduction as a protective coating in early 1960s, FBE coating fodicates, in Fusion bonded epoxy coatings the resin part is an "epoxy" type resin. "Epoxy" or "Oxirane" structure contains a three membered cyclic ring — one oxygen atom connected to two carbon atoms — in the resin molecule. This part is the most reactive group in the epoxy resins. Most commonly used FBE resins are derivatives of bisphenol A and epichlorohydrin. However, other types of resins (for example Bisphenol F type) are also commonly used in FBE formulations to achieve various properties, combinations or additions. Resins are also available in various molecular lengths, to provide unique properties to the final coating.

The second most important part of FBE coatings is the curing agent or hardener. Curing agents react either with the epoxy ring or with the hydroxyl groups, along the epoxy molecular chain.

Various types of curing agents, used in FBE manufacture, include dicyandiamide, aromatic amines, aliphatic diamines, etc. The selected curing agent determines the nature of the final FBE product — its cross-linking density, chemical resistance, brittleness, flexibility, etc. The ratio of epoxy resins and curing agents in a formulation is determined by their relative equivalent weights.

In addition to these two major components, FBE coatings include fillers, pigments, extenders and various additives, to provide desired properties. These components control characteristics such as permeability, hardness, colour, thickness, gouge resistance, etc. All of these components are normally dry solids, even though small quantities of liquid additives may be used in some FBE formulations. If used, these liquid components are sprayed into the formulation mix during preblending in the manufacturing process.

FBE Powder Manufacturing Process

Essential parts of a powder coating manufacturing plant are:

1. Weighting station,
2. Preblending station,
3. An extruder, and
4. A classifier or grinding unit.

The components of the FBE formulation are accurately weighed and are preblended in high speed mixers. The mix is then transferred to a high-shear extruder. FBE extruders incorporate a single or dual screw setup, rotating within a fixed clamshell barrel. A temperature gradient between 50°C to 100°C is used within the extruder barrel. This setup compresses the FBE blend, while heating and melting it to a semi-liquid form. During this process, the ingredients of the molten mix get dispersed thoroughly. Because of the fast operation of the extruder and relatively low temperature within the barrel, the epoxy and hardener components will not undergo a significant chemical reaction. The molten extrudate then passes between cold-rollers and becomes a solid, but highly brittle sheet. It further moves to a "Kibbler", which chops it to smaller chips. These chips are ground, using high speed grinders (classifiers) to a particle size of less than 150 micrometers (standard specifications requires 100 per cent pass through in 250 micrometer sieves and maximum 3 per cent retains in 150 micrometer sieve). The final product is packaged in closed containers — particular care is given to avoid moisture ingression. Normal storage temperatures of FBE powder coatings are below 25°C in air-conditioned warehouses.

FBE Coating Application Process

Regardless of the shape and type of steel surface to be coated, the FBE powder coating application has three essential stages:

1. The steel surface is cleaned to a high grade of cleaning,
2. The cleaned metal part is heated to the recommended FBE powder application temperature, and
3. The application and curing stage.

The advantage of pipe and rebar is that their round shape allows continuous linear application over the exterior surface, while the parts are moved in a linear conveyor through the powder application booth. This method ensures high production rates. On fittings, etc., the coating is applied by manual spraying guns. Another method of application is "fluid-dip" process, in which the heated components are dipped in a fluidised powder bed.

Surface Preparation — Blast Cleaning

"Blast cleaning" is the most commonly used method for preparation of steel surfaces. This method removes rust, scale, slats, etc., from the metal surface in an effective manner and produces an industrial grade cleaning and a "rough" surface finish. The roughness of the steel achieved after blasting is referred to as "profile", which is measured in micrometers or mils. Commonly used to profile ranges for FBE coatings are 37 to 100 micrometers (1.5 to 4 mils).

Profile increases the effective surface area of the steel. The cleanliness achieved is assessed as NACE grades, or in accordance with Swedish standard (SIS) terminology of "white-metal, near white-metal", etc.

It is very important to remove grease or oil contamination prior, to blast cleaning. Solvent cleaning, burn-off, etc., are commonly used for this purpose. In the blast cleaning process, compressed air (90 to 110 Ψ (psi)) is used to force an abrasive onto the surface, to be cleaned. Steel grit, steel shot, garnet, coal slag, etc., are the frequently used abrasive. Another method of blast cleaning is "centrifugal blast cleaning", which is especially used in cleaning the exterior of the pipes. In this method, abrasive is "thrown" to the rotating pipe body, using a specially designed wheel, which is rotated at high speed, while the abrasive is fed from the centre of the wheel.

Heating and FBE Powder Application

Heating can be achieved by using several methods, but the most commonly used ones are "induction heating" or "oven heating". The steel part is passed through a high frequency alternating current magnetic field, which heats the metal part to the required FBE coating application temperature. Other methods of heating are "oven heating", "infrared heating", etc. The FBE powder is placed on a "fluidisation bed". In a fluidisation bed, the powder particles are suspended in a stream of air, in which the powder will "behave" like a fluid. Once the air supply is turned off, the powder will remain in its original form. The fluidised powder is sprayed onto the hot substrate using suitable spray guns. An electrostatic spray gun incorporates an ioniser electrode on it, which gives the powder particles a positive electric charge. The steel to be coated is "grounded" through the conveyor. The charged powder particles uniform wraps around the substrate, and melts into a liquid form. Internal surfaces of pipes are coated using spraying lances, which travel from one end to the other end of the heated pipe at a uniform speed, while the pipe is being rotated in its longitudinal axis.

Standard coating thickness range of stand-alone FBE coatings is between 250 to 500 micrometers, even though lower or higher thickness ranges might be specified, depending on service conditions. The molten powder 'flows' into the profile and bonds with the steel. The molten powder will become a solid coating, when the 'gel time' is over, which usually occurs within few seconds after coating application. The resin part of coating will undergo cross-linking, which is known as "curing" under the hot condition. Complete curing is achieved either by the residual heat on the steel, or by the help of additional heating sources. Depending on the FBE coating system, full cure can be achieved in less than one minute to few minutes in case of long cure FBEs, which are used for internal pipe coating applications.

Rebars are coated in a similar manner as coating application, on the exterior of pipes. For FBE coating application on the interior of pipe surface, a lance is used. The lance enters into the pre-heated pipe, and starts spraying the powder from the opposite end, while the pipe is being rotated on its axis and the lance pulls out at a predetermined speed.

On fittings such as tees, elbows, bends, etc., powder can be sprayed using hand held spray guns. Small sized fittings can also be coated by dipping in a fluidised bed of powder, after heating the steel to the required powder application temperature. After field welding of the pipe ends, FBE can be applied on the weld area as well.

Advantages of FBE application over conventional liquid coating application are:

- Ease of application,
- Less waste of material,

- Rapid application,
- And cure schedules, which means faster production rates, and
- Finished coated pieces can be moved to the storage area within minutes after the application.

Polymer

A polymer is a substance composed of molecules with large molecular mass composed of repeating structural units, or monomers, connected by covalent chemical bonds. The term is derived from the Greek words: polys meaning many, and meros meaning parts. The individual molecules which comprise a polymer are referred to as polymer molecules, where the word "polymer" functions as an adjective. Well known examples of polymers include plastics and DNA.

While the term polymer in popular usage suggests "plastic", polymers comprise a large class of natural and synthetic materials with a variety of properties and purposes. Natural polymer materials such as shellac and amber have been in use for centuries. Paper is manufactured from cellulose, a naturally occurring polysaccharide found in plants. Biopolymers such as proteins and nucleic acids play important roles in biological processes.

Historical Development

The term polymer was coined in 1833 by Jons Jakob Berzelius. Around the same time Henri Braconnot did pioneering work in derivative cellulose compounds, perhaps the earliest important work in polymer science. The development of vulcanisation later in the nineteenth century improved the durability of the natural polymer rubber, signifying the first popularised semi-synthetic polymer. The first wholly synthetic polymer, Bakelite, was introduced in 1909.

Despite significant advances in synthesis and characterisation of polymers, a proper understanding of polymer molecular structure did not come until the 1920s.

Before that, scientists believed that polymers were clusters of small molecules (called colloids), without definite molecular weights, held together by an unknown force, a concept known as association theory. In 1922, Hermann Staudinger proposed that polymers consisted of long chains of atoms held together by covalent bonds, an idea which did not gain wide acceptance for over a decade, and for which Staudinger was ultimately awarded the Nobel Prize. In the intervening century, synthetic polymer materials such as Nylon, polyethylene, Teflon, and silicone have formed the basis for a burgeoning polymer industry.

Synthetic polymers today find application in nearly every industry and area of life. Polymers are widely used as adhesives and lubricants, as well as structural components for products ranging from childrens' toys to aircraft. Polymers such as poly(methyl methacrylate) find application as photoresist materials used in semi-conductor manufacturing and low-k dielectrics for use in high-performance microprocessors. Future applications include flexible polymer-based substrates for electronic displays and improved time-released and targeted drug delivery.

Polymer Science

Most polymer research may be categorised as polymer science, a subdiscipline of materials science which includes researchers in chemistry (especially organic chemistry), physics, and engineering. Polymer science may be roughly divided into two subdisciplines:

- Polymer chemistry or macromolecular chemistry, concerned with the chemical synthesis and chemical properties of polymers.
- Polymer physics, concerned with the bulk properties of polymer materials and engineering applications.

The field of polymer science is generally concerned with synthetic polymers, such as plastics, or chemical treatment and modification of natural polymers.

The study of biological polymers, their structure, function, and method of synthesis is generally the purview of biology, biochemistry, and biophysics. These disciplines share some of the terminology familiar to polymer science, especially when describing the synthesis of biopolymers such as DNA or polysaccharides. However, usage differences persist, such as the practice of using the term macromolecule to describe large non-polymer molecules and complexes of multiple molecular components, such as haemoglobin. Substances with distinct biological function are rarely described in the terminology of polymer science. For example, a protein is rarely referred to as a copolymer.

Polymer Synthesis

Polymers are synthesised by three primary methods: organic synthesis in a laboratory or factory, biological synthesis in living cells and organisms, or by chemical modification of naturally occurring polymers.

Organic Synthesis

In 1907, Leo Baekeland created the first completely synthetic polymer, Bakelite, by reacting phenol and formaldehyde at precisely controlled temperature and pressure. Subsequent work by Wallace Carothers in the 1920s demonstrated that polymers could be synthesised rationally from their constituent monomers. The intervening years have shown significant developments in rational polymer synthesis. Most commercially important polymers today are entirely synthetic and produced in high volume, on appropriately scaled organic synthetic techniques.

Laboratory synthetic methods are generally divided into two categories, condensation polymerisation and addition polymerisation. However, some newer methods such as plasma polymerisation do not fit neatly into either category. Synthetic polymerisation reactions may be carried out with or without a catalyst. Efforts towards rational synthesis of biopolymers via laboratory synthetic methods, especially artificial synthesis of proteins, is an area of intense research.

Biological Synthesis

Natural polymers and biopolymers formed in living cells may be synthesised by enzyme-mediated processes, such as the formation of DNA catalysed by DNA polymerase. The synthesis of proteins involves multiple enzyme-mediated processes to transcribe genetic information from the DNA and subsequently translate that information to synthesise the specified protein. The protein may be modified further following translation in order to provide appropriate structure and function.

Modification of Natural Polymers

Many commercially important polymers are synthesised by chemical modification of naturally occurring polymers. Prominent examples include the reaction of nitric acid and cellulose to form nitrocellulose and the formation of vulcanised rubber by heating natural rubber in the presence of sulphur.

Polymer Structure and Properties

Types of polymer 'properties' can be broadly divided into several categories based upon scale. At the nano-microscale are properties that directly describe the chain itself. These can be thought of as polymer structure. At an intermediate mesoscopic level are properties that describe the morphology of the polymer matrix in space. At the macroscopic level are properties that describe the bulk behaviour of the polymer.

Structure

The structural properties of a polymer relate to the physical arrangement of monomers along the backbone of the chain. Structure has a strong influence on the other properties of a polymer. For example, a linear chain polymer may be soluble or insoluble in water depending on whether it is composed of polar monomers (such as ethylene oxide) or non-polar monomers (such as styrene).

On the other hand, two samples of natural rubber may exhibit different durability even though their molecules comprise the same monomers. Polymer scientists have developed terminology to precisely describe both the nature of the monomers as well as their relative arrangement:

Monomer Identity

The identity of the monomers comprising the polymer is generally the first and most important attribute of a polymer. Polymer nomenclature is generally based upon the type of monomers comprising the polymer. Polymers that contain only a single type of monomer are known as homopolymers, while polymers containing a mixture of monomers are known as copolymers. Poly(styrene), for example, is composed only of styrene monomers, and is therefore is classified as a homopolymer. Ethylene-vinyl acetate, on the other hand, contains more than one variety of monomer and is thus a copolymer. Some biological polymers are composed of a variety of different but structurally related monomers, such as polynucleotides composed of nucleotide subunits.

A polymer molecule containing ionisable subunits is known as a polyelectrolyte. An ionomer is a subclass of polyelectrolyte with a low fraction of ionisable subunits.

Chain Linearity

The simplest form of polymer molecule is a straight chain or linear polymer, composed of a single main chain. The flexibility of an unbranched chain polymer is characterised by its persistence length. A branched polymer molecule is composed of a main chain with one or more substituent side chains or branches. Special types of branched polymers include star polymers, comb polymers, and brush polymers. If the polymer contains a side chain that has a different composition or configuration than the main chain the polymer is called a graft or grafted polymer. A cross-link suggests a branch point from which four or more distinct chains emanate. A polymer molecule with a high degree of cross-linking is referred to as a polymer network. Sufficiently high cross-link concentrations may lead to the formation of an 'infinite network', also known as a 'gel', in which networks of chains are of unlimited extend—there is essentially all chains have linked into one molecule.

Chain Size

Polymer bulk properties may be strongly dependent on the size of the polymer chain. Like any molecule, a polymer molecule's size may be described in terms of molecular weight or mass. In polymers, however, the molecular mass may be expressed in terms of degree of polymerisation, essentially the number of monomer units which comprise the polymer. For synthetic polymers, the molecular weight is expressed statistically to describe the distribution of molecular weights in the sample. This is because of the fact that almost all industrial processes produce a distribution of polymer chain sizes. Examples of such statistics include the number average molecular weight and weight average molecular weight. The ratio of these two values is the polydispersity index, commonly used to express the "width" of the molecular weight.

The space occupied by a polymer molecule is generally expressed in terms of radius of gyration or excluded volume.

Monomer Arrangement in Copolymers

Monomers within a copolymer may be organised along the backbone in a variety of ways.

- Alternating copolymers possess regularly alternating monomer residues.
- Periodic copolymers have monomer residue types arranged in a repeating sequence.
- Random copolymers have a random sequence of monomer residue types.
- Statistical copolymers have monomer residues arranged according to a known statistical rule.
- Block copolymers have two or more homopolymer subunits linked by covalent bonds. Block copolymers with two or three distinct blocks are called diblock copolymers and triblock copolymers, respectively.

Tacticity

This property relates to how neighbouring structural units within the polymer are oriented with respect to one another.

Morphological Properties

Crystallinity: When applied to polymers, the term crystalline has a somewhat ambiguous usage. In some cases, the term crystalline finds identical usage to that used in conventional crystallography.

For example, the structure of a crystalline protein or polynucleotide, such as a sample prepared for x-ray crystallography, may be defined in terms of a conventional unit cell composed of one or more polymer molecules with cell dimensions of hundreds of angstroms or more.

A synthetic polymer may be described as crystalline if it contains regions of three-dimensional ordering on atomic (rather than macromolecular) length scales, usually arising from intramolecular folding and/or stacking of adjacent chains. Synthetic polymers may consist of both crystalline and amorphous regions; the degree of crystallinity may be expressed in terms of a weight fraction or volume fraction of crystalline material. Few synthetic polymers are entirely crystalline.

Bulk Properties

The bulk properties of a polymer are those most often of end-use interest. These are the properties that dictate how the polymer actually behaves on a macroscopic scale.

Tensile strength

The tensile strength of a material quantifies how much stress the material will endure before failing. This is very important in applications that rely upon polymer's physical strength or durability. For example, a rubber band with a higher tensile strength will hold a greater weigh before snapping. In general tensile strength increases with polymer chain length.

Young's Modulus of Elasticity

This parameter quantifies the elasticity of the polymer. It is defined, for small strains, as the ratio of rate of change of stress to strain. Like tensile strength this is highly relevant in polymer applications involving the physical properties of polymers, such as rubber bands.

Transport Properties

Transport properties such as diffusivity relate to how rapidly molecules move through the polymer matrix. These are very important in many applications of polymers for films and membranes.

Pure Component Phase Behaviour

Melting Point: The term "melting point" when applied to polymers suggests not a solid-liquid phase transition but a transition from a crystalline or semi-crystalline phase to a solid amorphous phase. Though abbreviated as simply "T_m", the property in question is more properly called the "crystalline melting temperature". Among synthetic polymers, crystalline melting is only discussed with regards to thermoplastics, as thermosetting polymers will decompose at high temperatures rather than melt.

Boiling Point

The boiling point of a polymer substance is never defined due to the fact that polymers will decompose before reaching theoretical boiling temperatures.

Glass Transition Temperature (T_g)

A parameter of particular interest in synthetic polymer manufacturing is the glass transition temperature (T_g), which describes the temperature at which amorphous polymers undergo a second order phase transition from a rubbery, viscous amorphous solid to a brittle, glassy amorphous solid. The glass transition temperature may be engineered by altering the degree of branching or cross-linking in the polymer or by the addition of plasticiser.

Polymer Solution Behaviour

In general, polymeric mixtures are far less miscible than mixtures of small molecule materials. This effect is a result of the fact that the driving force for mixing is usually entropics, not energetics. In other words, miscible materials usually form a solution not because their interaction with each other is more favourable than their self-interaction but because of an increase in entropy and hence free energy associated with increasing the amount of volume available to each component. This increase in entropy scales with the number of particles (or moles) being mixed.

Since polymeric molecules are much larger and hence generally have much higher specific volumes than small molecules, the number of molecules involved in a polymeric mixture are far less than the number in a small molecule mixture of equal volume.

The energetics of mixing, on the other hand, are comparable on a per volume basis for polymeric and small molecule mixtures. This tends to increase the free energy of mixing for polymer solutions and thus make solvation less favourable. Thus, concentrated solutions of polymers are far rarer than those of small molecules.

In dilute solution, the properties of the polymer are characterised by the interaction between the solvent and the polymer. In a *good solvent*, the polymer appears swollen and occupies a large volume. In this scenario, intermolecular forces between the solvent and monomer subunits dominate over intramolecular interactions. In a bad solvent or poor solvent, intramolecular forces dominate and the chain contracts. In the *theta solvent*, or the state of the polymer solution where the value of the second virial coefficient becomes 0, the intermolecular polymer-solvent repulsion balances exactly the intramolecular monomer-monomer attraction. Under the theta condition (also called the Flory condition) the polymer behaves like an ideal random coil.

Polymer Structure/Property Relationships

Polymer bulk properties are strongly dependent upon their structure and mesoscopic behaviour. A number of qualitative relationships between structure and properties are known.

Chain Length

Increasing chain length tends to decrease chain mobility, increase strength and toughness, and increase the glass transition temperature (T_g). This is a result of the increase in chain interactions such as Van der Waals attractions and entanglements that come with increased chain length. These interactions tend to fix the individual chains more strongly in position and resist deformations and matrix breakup, both at higher stresses and higher temperatures.

Branching

Branching of polymer chains also affect the bulk properties of polymers. Long chain branches may increase polymer strength, toughness, and Tg due to an increase in the number of entanglements per chain. Random length and atactic short chains, on the other hand, may reduce polymer strength due to disruption of organisation.

Short side chains may likewise reduce crystallinity due to disruption of the crystal structure. Reduced crystallinity may also be associated with increased transparency due to light scattering by small crystalline regions.

A good example of this effect is related to the range of physical attributes of polyethylene. High density polyethylene (HDPE) has a very low degree of branching, is quite stiff, and is used in applications such as milk jugs. Low density polyethylene (LDPE), on the other hand, has significant numbers of short branches, is quite flexible, and is used in applications such as plastic films. The branching index of the polymer is a parameter that characterises the effect of long-chain branches on the size of a branched macromolecule in solution.

Chemical Cross-linking

Cross-linking tends to increase T_g and increase strength and toughness. Cross-linking consists of the formation of chemical bonds between chains.

Among other applications, this process is used to strengthen rubbers in a process known as Vulcanisation, which is based on cross-linking by sulphur. Car tires, for example, are highly cross-linked in order to reduce permeation of air out of the tire and toughen the tire durability. Eraser rubber, on the other hand, is not cross-linked to allow flaking of the rubber and prevent damage to the paper.

Inclusion of Plasticisers

Inclusion of Plasticisers tends to lower T_g and increase polymer flexibility. Plasticisers are generally small molecules that are chemically similar to the polymer and create gaps between polymer chains for greater mobility and reduced interchain interactions. A good example of the action of plasticisers is related to polyvinylchlorides or PVCs. A uPVC or unplastiscised polyvinylchloride is used for things such as pipes. A pipe has no plasticisers in it because it needs to remain strong and heat resistant. Plasticised PVC is used for clothing for a flexible quality. Plasticisers are also put in some types of cling film to make the polymer more flexible.

Degree of Crystallinity

Increasing degree of crystallinity tends to make a polymer more rigid. It can also lead to greater brittlness. Polymers with degree of crystallinity approaching zero or one will tend to be transparent, while polymers with intermediate degrees of crystallinity will tend to be opaque due to light scattering by crystalline / glassy regions.

Standardised Polymer Nomenclature

There are multiple conventions for naming polymer substances. Many commonly used polymers, such as those found in consumer products, are referred to by a *common* or *trivial* name.

The trivial name is assigned based on historical precedent or popular usage rather than a standardised naming convention. Both the American Chemical Society and IUPAC have proposed standardised naming conventions.

In both standardised conventions the polymers names are intended to reflect the monomer(s) from which they are synthesised rather than the precise nature of the repeating subunit.

For example, the polymer synthesised from the simple alkene ethene is called polyethylene, retaining the -ene suffix even though the double bond is removed during the polymerisation process:

$$\left(H_2C{=}CH_2 \right)_n \xrightarrow[\text{polymerisation}]{\text{radical addition}} \text{—}CH_2\text{—}CH_2\text{—}CH_2\text{—}CH_2\text{—}CH_2\text{—}CH_2\text{—}$$

The Polymerisation of ethene in to poly(ethene)

$$\left(\text{—}CH_2\text{—}CH_2\text{—} \right)_n$$

Chemical Properties of Polymers

The attractive forces between polymer chains play a large part in determining a polymer's properties. Because polymer chains are so long, these interchain forces are amplified far beyond the attractions between conventional molecules. Different side groups on the polymer can lend the polymer to ionic bonding or hydrogen bonding between its own chains. These stronger forces typically result in higher tensile strength and melting points.

The intermolecular forces in polymers can be affected by dipoles in the monomer units. Polymers containing amide or carbonyl groups can form hydrogen bonds between adjacent chains; the partially positively charged hydrogen atoms in N-H groups of one chain are strongly attracted to the partially negatively charged oxygen atoms in C=O groups on another. These strong hydrogen bonds, for example, result in the high tensile strength and melting point of polymers containing urethane or urea linkages. Polyesters have dipole-dipole bonding between the oxygen atoms in C=O groups and the hydrogen atoms in H-C groups. Dipole bonding is not as strong as hydrogen bonding, so a polyester's melting point and strength are lower than Kevlar's, but polyesters have greater flexibility.

Ethene, however, has no permanent dipole. The attractive forces between polyethylene chains arise from weak van der Waals forces. Molecules can be thought of as being surrounded by a cloud of negative electrons.

As two polymer chains approach, their electron clouds repel one another. This has the effect of lowering the electron density on one side of a polymer chain, creating a slight positive dipole on this side. This charge is enough to actually attract the second polymer chain. Van der Waals forces are quite weak, however, so polyethene can have a lower melting temperature compared to other polymers.

Polymer Characterisation

The characterisation of a polymer requires several parameters which need to be specified. This is because a polymer actually consists of a statistical distribution of chains of varying lengths, and each chain consists of monomer residues which affect its properties.

A variety of lab techniques are used to determine the properties of polymers. Techniques such as wide angle X-ray scattering, small angle X-ray scattering, and small angle neutron scattering are used to determine the crystalline structure of polymers. Gel permeation chromatography is used to determine the number average molecular weight, weight average molecular weight, and polydispersity. FTIR, Raman and NMR can be used to determine composition. Thermal properties such as the glass transition temperature and melting point can be determined by differential scanning colorimetry and dynamic mechanical analysis. Pyrolysis followed by analysis of the fragments is one more technique for determining the possible structure of the polymer.

Polymer Degradation

Polymer degradation is a change in the properties—tensile strength, colour, shape, etc.—of a polymer or polymer based product under the influence of one or more environmental factors such as heat, light or chemicals. It is often due to the hydrolysis of the bonds connecting the polymer chain, which in turn leads to a decrease in the molecular mass of the polymer. These changes may be undesirable, such as changes during use, or desirable, as in biodegradation or deliberately lowering the molecular mass of a polymer. Such changes occur primarily because of the effect of these factors on the chemical composition of the polymer.

The degradation of polymers to form smaller moleculars may proceed by random scission or specific scission. The degradation of polyethylene occurs by *random scission*—that is by a random breakage of the linkages (bonds) that hold the atoms of the polymer together. When heated above 450 Celsius it degrades to form a mixture of hydrocarbons. Other polymers-like polyalphamethylstyrene—undergo 'specific' chain scission with breakage occurring only at the ends. They literally unzip or depolymerise to become the constituent monomer.

In a finished product such a change is to be prevented or delayed. However the degradation process can be useful from the view points of understanding the structure of a polymer or recycling/reusing the polymer waste to prevent or reduce environmental pollution.

Polylactic acid and Polyglycolic acid, for example, are two polymers that are useful for their ability to degrade under aqueous conditions. A copolymer of these polymers is used for biomedical applications such as hydrolysable stitches that degrade over time after they are applied to a wound. These materials can also be used for plastics that will degrade over time after they are used and will therefore not remain as litter.

Industry

Today there are primarily six commodity polymers in use, namely polyethylene, polypropylene, polyvinyl chloride, polyethylene terephthalate, polystyrene and polycarbonate. These make up nearly 98 per cent of all polymers and plastics encountered in daily life.

Each of these polymers has its own characteristic modes of degradation and resistances to heat, light and chemicals.

Cracking of Polymers

Cracking is the process by which a polymer is divided into its subcomponents or monomers. The resulting subcomponents are more viscous than the original polymer.

Sputter Deposition

Sputter deposition is a physical vapour deposition PVD method of depositing thin films by sputtering a block of source material onto a "substrate".

Sputtered atoms ejected into the gas phase are not in their thermodynamic equilibrium state, and tend to deposit on all surfaces in the vacuum chamber. A substrate (such as a wafer) placed in the chamber will be coated with a thin film. Sputtering usually uses an argon plasma.

Sputtering is used extensively in the semi-conductor industry to deposit thin films of various materials in integrated circuit processing. Thin antireflection coatings on glass for optical applications are also deposited by sputtering. Because of the low substrate temperatures used, sputtering is an ideal method to deposit contact metals for thin-film transistors. Perhaps the most familiar products of sputtering are low-emissivity coatings on glass, used in double-pane window assemblies. The coating is a multilayer containing silver and metal oxides such as zinc oxide, tin oxide, or titanium dioxide.

Sputtering is also used as the process to deposit the metal (Aluminium) layer during the fabrication of CD and DVD discs.

Comparison with other Deposition Methods

One important advantage of sputtering as a deposition technique is that the deposited films have the same composition as the source material. The equality of the film and target stoichiometry might be surprising since the sputter yield depends on the atomic weight of the atoms in the target. One might therefore expect one component of an alloy or mixture to sputter faster than the other components, leading to an enrichment of that component in the deposit.

However, since only surface atoms can be sputtered, the faster ejection of one element leaves the surface enriched with the others, effectively counteracting the difference in sputter rates. This is in contrast to thermal evaporation techniques, where one component of the source may have a higher vapour pressure, resulting in a deposited film with a different composition than the source.

Sputter deposition also has an advantage over molecular beam epitaxy (MBE) due to its speed. The higher rate of deposition results in lower impurity incorporation because fewer impurities are able to reach the surface of the substrate in the same amount of time. Sputtering methods are consequently able to use process gases with far higher impurity concentrations than the vacuum pressure that MBE methods can tolerate. During sputter deposition the substrate may be bombarded by energetic ions and neutral atoms. Ions can be deflected with a substrate bias and neutral bombardment can be minimised by off-axis sputtering, but only at a cost in deposition rate. Plastic substrates cannot tolerate the bombardment and are usually coated via evaporation.

Types of Sputter Deposition

Sputtering sources are usually magnetrons that utilise strong electric and magnetic fields to trap electrons close to the surface of the magnetron, which is known as the target.

The electrons follow helical paths around the magnetic field lines undergoing more ionising collisions with gaseous neutrals near the target surface than would otherwise occur. The sputter gas is inert, typically argon. The extra argon ions created as a result of these collisions leads to a higher deposition rate. It also means that the plasma can be sustained at a lower pressure. The sputtered atoms are neutrally charged and so are unaffected by the magnetic trap.

Charge build-up on insulating targets can be avoided with the use of RF sputtering where the sign of the anode-cathode bias is varied at a high rate. RF sputtering works well to produce highly insulating oxide films but only with the added expense of RF power supplies and impedance matching networks. Stray magnetic fields leaking from ferromagnetic targets also disturb the sputtering process. Specially designed sputter guns with unusually strong permanent magnets must often be used in compensation.

Ion-beam Sputtering

Ion-beam sputtering (IBS) is a method in which the target is external to the ion source. A source can work without any magnetic field like in a Hot filament ionisation gauge. In a Kaufman source ions are generated by collisions with electrons that are confined by a magnetic field as in a magnetron. They are then accelerated by the electric field emanating from a grid towards a target.

As the ions leave the source they are neutralised by electrons from a second external filament. IBS has an advantage in that the energy and flux of ions can be controlled independently. Since the flux that strikes the target is composed of neutral atoms, either insulating or conducting targets can be sputtered. IBS has found application in the manufacture of thin-film heads for disk drives.

A pressure gradient between the ion source and the sample chamber is generated by placing the gas inlet at the source and shooting through a tube in into the sample chamber. This saves gas and reduces contamination in UHV applications. The principal drawback of IBS is the large amount of maintenance required to keep the ion source operating.

Reactive Sputtering

In reactive sputtering, the deposited film is formed by chemical reaction between the target material and a gas which is introduced into the vacuum chamber. Oxide and nitride films are often fabricated using reactive sputtering.

The composition of the film can be controlled by varying the relative pressures of the inert and reactive gases. Film stoichiometry is an important parameter for optimising functional properties like the stress in SiNx and the index of refraction of SiOx. The transparent indium tin oxide conductor that is used in optoelectronics and solar cells is made by reactive sputtering.

Ion-assisted Deposition

In ion-assisted deposition (IAD), the substrate is exposed to a secondary ion beam operating at a lower power than the sputter gun. Usually a Kaufman source like that used in IBS supplies the secondary beam. IAD can be used to deposit carbon in diamond-like form on a substrate.

Any carbon atoms landing on the substrate which fail to bond properly in the diamond crystal lattice will be knocked off by the secondary beam. NASA used this technique to experiment with depositing diamond films on turbine blades in the 1980s. IAS is used in other important industrial applications such as creating tetrahedral amorphous carbon surface coatings on hard disk platters and hard transition metal nitride coatings on medical implants.

High-target-utilisation Sputtering

High-target-utilisation sputtering "HiTUS" is specialised commercial process. The process based upon the remote generation of a high density plasma. The plasma is generated in a side chamber opening into the main process chamber, containing the target and the substrate to be coated. As the plasma is generated remotely, and not from the target itself (as in conventional magnetron sputtering), the ion current to the target is independent of the voltage applied to the target.

High Power Impulse Magnetron Sputtering (HIPIMS)

HIPIMS is a method for physical vapour deposition of thin films which is based on magnetron sputter deposition. HIPIMS utilises extremely high power densities of the order of kWcm^2 in short pulses (impulses) of tens of microseconds at low duty cycle of < 10 per cent.

5

Industrial-Chemical Analysis

Iodometry

Iodometry is a method of volumetric chemical analysis, a titration where the appearance or disappearance of elementary iodine indicates the end point. Usual reagents are sodium thiosulphate as titrant, starch as an indicator (it forms blue complex with iodine molecules—though polyvinyl alcohol has started to be used recently as well), and an iodine compound (iodide or iodate, depending on the desired reaction with the sample).

The principal reaction is the reduction of iodine to iodide by thiosulphate:

$$I_2 + 2\,S_2O_3^{2-} \longrightarrow S_4O_6^{2-} + 2\,I^-$$

A common and illustrative use of iodometry is the measurement of concentration of chlorine in water. Chlorine in pH under 8 oxidises iodide to iodine. An overabundance of potassium iodide is added to the known amount of sample in acidic environment (pH < 4, the reaction is not complete in more alkaline pH). Starch is added, forming blue clathrate complex with the liberated iodine. The blue solution is then titrated with thiosulphate until the blue colour vanishes.

Two possible sources of error can influence the outcome of the iodometric titration. One is the air oxidation of acid-iodide solution and the other is the volatility of I_2. The first one can be eliminated by adding an excess of sodium carbonate in the reaction vessel. This removes oxygen in the vessel by forming carbon dioxide (which is heavier than air). The other error can be reduced by using an excess of iodide solution which captures liberated iodine to form triiodide ions, I_3^-.

Redox

Redox (shorthand for reduction/oxidation reaction) describes all chemical reactions in which atoms have their oxidation number (oxidation state) changed.

This can be a simple redox process, such as the oxidation of carbon to yield carbon dioxide, it could be the reduction of carbon by hydrogen to yield methane (CH_4), or a complex process such as the oxidation of sugar in the human body, through a series of very complex electron transfer processes.

The term redox comes from the two concepts of reduction and oxidation. It can be explained in simple terms:

- Oxidation describes the loss of electrons by a molecule, atom or ion.
- Reduction describes the gain of electrons by a molecule, atom or ion However, these descriptions though sufficient for many purposes) are not truly correct. Oxidation and reduction properly refer to a change in oxidation number — the actual transfer of electrons may never occur. Thus, oxidation is better defined as an increase in oxidation number, and reduction as a decrease in oxidation number. In practice, the transfer of electrons will always cause a change in oxidation number, but there are many reactions which are classed as "redox" even though no electron transfer occurs (such as those involving covalent bonds).

Non-redox reactions, which do not involve changes in formal charge, are known as metathesis reactions.

Oxidising and Reducing Agents

Substances that have the ability to oxidise other substances are said to be oxidative and are known as oxidising agents, oxidants or oxidisers.

Put in another way, the oxidant removes electrons from another substance, and is thus reduced itself. And because it "accepts" electrons it is also called an electron acceptor. Oxidants are usually chemical substances with elements in high oxidation numbers (e.g., H_2O_2, MnO_4^-, CrO_3, $Cr_2O_7^{2-}$, OsO_4) or highly electronegative substances that can gain one or two extra electrons by oxidising a substance (O, F, Cl, Br).

Substances that have the ability to reduce other substances are said to be reductive and are known as reducing agents, reductants, or reducers. Put in another way, the reductant transfers electrons to another substance, and is thus oxidised itself. And because it "donates" electrons it is also called an electron donor. Reductants in chemistry are very diverse.

Metal reduction—electropositive elemental metals can be used (Li, Na, Mg, Fe, Zn, Al.). These metals donate or give away electrons readily. Other kinds of reductants are hydride transfer reagents ($NaBH_4$, $LiAlH_4$), these reagents are widely used in organic chemistry, primarily in the reduction of carbonyl compounds to alcohols.

Another useful method is reductions involving hydrogen gas (H_2) with a palladium, platinum, or nickel catalyst. These catalytic reductions are primarily used in the reduction of carbon-carbon double or triple bonds.

The chemical way to look at redox processes is that the reductant transfers electrons to the oxidant. Thus, in the reaction, the reductant or reducing agent loses electrons and is oxidised and the oxidant or oxidising agent gains electrons and is reduced.

Light and heat speed up the movement of molecules and that therefore increases the speed of which electrons are lost by molecules, atoms or ions(oxidation).

Oxidation in Industry

Oxidation is used in a wide variety of industries such as in the production of cleaning products.

Redox reactions are the foundation of electrochemical cells.

Examples of Eedox Eeactions

A good example is the reaction between hydrogen and fluorine:

$$H_2 + F_2 \longrightarrow 2HF$$

We can write this overall reaction as two half-reactions: the oxidation reaction

$$H_2 \longrightarrow 2H^+ + 2e^-$$

and the reduction reaction:

$$F_2 + 2e^- \longrightarrow 2F^-$$

Analysing each half-reaction in isolation can often make the overall chemical process clearer. Because there is no net change in charge during a redox reaction, the number of electrons in excess in the oxidation reaction must equal the number consumed by the reduction reaction.

Elements, even in molecular form, always have an oxidation number of zero.

In the first half reaction, hydrogen is oxidised from an oxidation number of zero to an oxidation number of +1. In the second half reaction, fluorine is reduced from an oxidation number of zero to an oxidation number of "1.

When adding the reactions together the electrons cancel:

$$H_2 \longrightarrow 2H^+ + 2e^-$$

$$F_2 + 2e^- \longrightarrow 2F^-$$

$$H_2 + F_2 \longrightarrow 2H^+ + 2F^-$$

And the ions combine to form hydrogen fluoride:

$$H_2 + F_2 \longrightarrow 2H^+ + 2F^- \longrightarrow 2HF$$

Other Examples

- Iron (II) oxidises to iron (III):

  $Fe^{2+} \longrightarrow Fe^{3+} + e^-$

- Hydrogen peroxide reduces to hydroxide in the presence of an acid:

  $H_2O_2 + 2\ e^- \longrightarrow 2\ OH^-$

  overall equation for the above:

  $2Fe^{2+} + H_2O_2 + 2H^+ \longrightarrow 2Fe^{3+} + 2H_2O$

- Denitrification, nitrate reduces to nitrogen in the presence of an acid:

  $2NO_3^- + 10e^- + 12\ H^+ \longrightarrow N_2 + 6H_2O$

- Iron oxidises to iron (III) oxide and oxygen is reduced forming iron (III) oxide (commonly known as rusting, which is similar to tarnishing):

  $4Fe + 3O_2 \longrightarrow 2\ Fe_2O_3$

- Combustion of hydrocarbons, e.g. in an internal combustion engine, produces water, carbon dioxide, some partially oxidised forms such as carbon monoxide and heat energy. Complete oxidation of materials containing carbon produces carbon dioxide.
- In organic chemistry, stepwise oxidation of a hydrocarbon produces water and, successively, an alcohol, an aldehyde or a ketone, carboxylic acid, and then a peroxide.
- In biology many important processes involve redox reactions. Cell respiration, for instance, is the oxidation of glucose ($C_6H_{12}O_6$) to CO_2 and the reduction of oxygen to water. The summary equation for cell respiration is:

  $C_6H_{12}O_6 + 6\ O_2 \longrightarrow 6\ CO_2 + 6\ H_2O$

The process of cell respiration also depends heavily on the reduction of NAD^+ to NADH and the reverse reaction (the oxidation of NADH to NAD^+). Photosynthesis is essentially the reverse of the redox reaction in cell respiration:

$$6\ CO_2 + 6\ H_2O + \text{light energy} \longrightarrow C_6H_{12}O_6 + 6\ H_2O$$

Redox Reactions in Biology

Much biological energy is stored and released by means of redox reactions. Photosynthesis involves the reduction of carbon dioxide into sugars and the oxidation of water into molecular oxygen.

The reverse reaction, respiration, oxidises sugars to produce carbon dioxide and water. As intermediate steps, the reduced carbon compounds are used to reduce nicotinamide adenine dinucleotide (NAD^+), which then contributes to the creation of a proton gradient, which drives the synthesis of adenosine triphosphate (ATP) and is maintained by the reduction of oxygen. In animal cells, mitochondria perform similar functions.

The term redox state is often used to describe the balance of NAD^+/NADH and $NADP^+$/NADPH in a biological system such as a cell or organ.

The redox state is reflected in the balance of several sets of metabolites (e.g., lactate and pyruvate, beta-hydroxybutyrate and acetoacetate) whose interconversion is dependent on these ratios. An abnormal redox state can develop in a variety of deleterious situations, such as hypoxia, shock, and sepsis. Redox signalling involves the control of cellular processes by redox processes.

Mnemonics

The key terms involved in redox can be confusing. For example, an element that is oxidised loses electrons; however, that element is referred to as the reducing agent. Likewise, an element that is reduced gains electrons and is referred to as the oxidising agent. Several acronyms and mnemonics are often used to remember what is happening:

- *"OIL RIG"*: Oxidation Is Loss, Reduction Is Gain.
- *"LEO the lion says GER"*: Losing Electrons is Oxidation, Gaining Electrons is Reduction.
- *"LEOA the lion's girlfriend says GERC"*: Loss of Electrons is Oxidation at Anode, Gain of Electrons is Reduction at Cathode.

- "EOH" reminds you to include an Electron and Oxygen and Hydrogen atom(s) in your complex equations.
- *"GEROA", "LEORA":* Gain Electron Reduction Oxidising Agent, Lose Electron Oxidation Reducing Agent
- *"GAWROL":* GAiner Was Reduced, by Oxidised Loser.

Another example of remembering loss/gain of electrons is that oxidation is the loss of electrons, giving emphasis to the 'o'. This emphasis of 'o' can be remembered as Oxidation

Redox Cycling

A wide variety of aromatic compounds are enzymatically reduced to form free radicals that contain one more electron than their parent compounds. In general, the electron donor is any of a wide variety of flavoenzymes and their co-enzymes. Once formed, these anion free radicals reduce molecular oxygen to superoxide and regenerate the unchanged parent compound. The net reaction is the oxidation of the flavoenzyme's co-enzymes and the reduction of molecular oxygen to form superoxide. This catalytic behaviour has been described as futile cycle or redox cycling.

Examples of redox cycling-inducing molecules are the herbicide paraquat and other viologens and quinones such as menadione.

Bessemer Process

The Bessemer process was the first inexpensive industrial process for the mass-production of steel from molten pig iron. The process is named after its inventor, Henry Bessemer, who took out a patent on the process in 1855. The process was independently discovered in 1851 by William Kelly. The process had also been used outside of Europe for hundreds of years, but not on an industrial scale. The key principle is removal of impurities from the iron by oxidation through air being blown through the molten iron. The oxidation also raises the temperature of the iron mass and keeps it molten.

The Details of the Process

Bessemer Converter: The process is carried on in a large ovoid steel container lined with clay or dolomite called the Bessemer converter. The capacity of a converter was from 8 to 30 tons of molten iron with a usual charge being around 15 tons. At the top of the converter is an opening, usually tilted to the side relative to the body of the vessel, through which the iron is introduced and the finished product removed. The bottom is perforated with a number of channels called tuyères through which air is forced into the converter. The converter is pivoted on trunnions so that it can be rotated to receive the charge, turned upright during conversion, and then rotated again for pouring out the molten steel at the end.

Oxidation

The oxidation process removes impurities such as silicon, manganese, and carbon as oxides. These oxides either escape as gas or form a solid slag. The refractory lining of the converter also plays a role in the conversion—the clay lining is used in the *acid Bessemer*, in which there is low phosphorus in the raw material. Dolomite is used when the phosphorus content is high in the *basic Bessemer* (limestone or magnesite linings are also sometimes used instead of dolomite)—this is also known as a Gilchrist-Thomas converter. In order to give the steel the desired properties, other substances could be added to the molten steel when conversion was complete, such as spiegeleisen (an iron-carbon-manganese alloy).

Managing the Process

When the required steel had been formed, it was poured out into ladles and then transferred into moulds and the lighter slag is left behind. The conversion process (called the "blow") was completed in around twenty minutes. During this period the progress of the oxidation of the impurities was judged by the appearance of the flame issuing from the mouth of the converter: the modern use of photoelectric methods of recording the characteristics of the flame has greatly aided the blower in controlling the final quality of the product. After the blow, the liquid metal was recarburised to the desired point and other alloying materials are added, depending on the desired product.

Predecessor Processes

Before the Bessemer process Britain had no practical method of reducing the carbon content of pig iron. Steel was manufactured by the reverse process of adding carbon to carbon-free wrought iron, usually imported from Sweden. The manufacturing process, called cementation process, consisted of heating bars of wrought iron together with charcoal for periods of up to a week in a long stone box. This produced blister steel. Up to 3 tons of expensive coke was burnt for each ton of steel produced. Such steel when rolled into bars was sold at £50 to £60 a long ton. The most difficult and work-intensive part of the process was however the production of wrought iron done in finery forges in Sweden.

This process was refined in the 1700s with the introduction of Benjamin Huntsman's crucible steel making technique, which added an additional three hours firing time, and additional massive quantities of coke. In making crucible steel, the blister steel bars were broken into pieces and melted in small crucibles each containing 20 kg or so. This produced higher quality crucible steel, and increased the cost. The Bessemer process reduced to about ½ hour the time to make steel of this quality, while requiring only the coke needed initially to melt the pig iron. The earliest Bessemer converters produced steel for £7 a long ton, although they priced it initially at around £40 a ton.

History

Both Bessemer and Huntsman were based in the city of Sheffield, England. Sheffield has an international reputation for steel-making, which dates from 1740, when Benjamin Huntsman discovered the crucible technique for steel manufacture, at his workshop in the district of Handsworth. This process had an enormous impact on the quantity and quality of steel production and was only made obsolete, a century later, in 1856 by Henry Bessemer's invention of the Bessemer converter which allowed the true mass production of steel. Bessemer had moved his Bessemer Steel Company to Sheffield to be at the heart of the industry. The city's Kelham Island Museum still maintains one of the UK's last examples of a working Bessemer converter [from Workington, Cumbria] for public viewing.

Importance

The Bessemer process revolutionised the world. Prior to its widespread use, steel was far too expensive to use in most applications, and wrought iron was used throughout the Industrial Revolution. After its introduction, steel and wrought iron were similarly priced, and all manufacture turned to steel.

Obsolescence

In the US commercial steel production using this method stopped in 1968. It was replaced by processes such as Linz-Donawitz process that offered better control of final chemistry. The Bessemer process was so fast (10-20 minutes for a heat) that it allowed little time for chemical analysis or adjustment of the alloying elements in the steel.

Bessemer converters did not remove phosphorus efficiently from the molten steel; as low-phosphorous ores became more expensive, conversion costs increased. The process only permitted a limited amount of scrap steel to be charged, further increasing costs, especially when scrap was inexpensive. Certain grades of steel were sensitive to the nitrogen which was part of the air blast passing through the steel.

Oxidative Addition

Oxidative addition and reductive elimination are two important classes of reactions in organometallic chemistry. Their relationship is shown below where y represents the number of ligands on the metal and n is the oxidation state of the metal.

In oxidative addition, a metal complex with vacant coordination sites and a relatively low oxidation state is oxidised by the insertion of the metal into a covalent bond (X-Y). Both the formal oxidation state of the metal, n, and the electron count of the complex increase by two. Although oxidative additions can occur with the insertion of a metal into many different covalent bonds, they are most commonly seen with H-H and carbon(sp^3)-halogen bonds. Carbon that is sp^2 hybridised, as in the case of a vinyl group, can also undergo oxidative addition. This process proceeds with retention of configuration about the double bond.

The reverse of oxidative addition is reductive elimination. Reductive elimination is favoured when the newly formed X-Y bond is strong. For reductive elimination to occur the two groups (X and Y) should be adjacent to each other in the metal's coordination sphere.

An example of oxidative addition is the reaction of Vaska's complex, *trans*-$IrCl(CO)[P(C_6H_5)_3]_2$, with hydrogen. In this transformation, the metal oxidation state changes from Ir(I) to Ir(III) because the product is described as Ir^{3+} bound to three anions: Cl^-, and two hydride, H^-, ligands. As shown below, the metal complex initially has 16 valence electrons and a coordination number of four. After the addition of hydrogen, the complex has 18 electrons and a coordination number of six. The reaction can proceed in the opposite direction. In this case hydrogen gas would be lost and the metal complex would be reduced. This backwards reaction is an example of reductive elimination.

$$(C_6H_5)_3P\text{—}Ir(Cl)(OC)\text{—}P(C_6H_5)_3 \underset{}{\overset{H_2}{\rightleftharpoons}} (C_6H_5)_3P\text{—}Ir(H)(H)(OC)(Cl)\text{—}P(C_6H_5)_3$$

16e⁻ 18e⁻

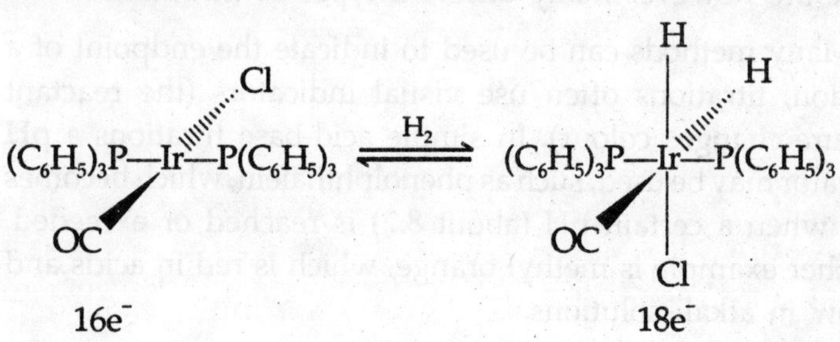

Just as a metal oxidatively inserts itself into a H-H bond, it can also oxidatively add to C-H bonds. This process is called C-H bond activation and is an active research area because of its potential value in converting petroleum-derived hydrocarbons into more complex products.

Oxidative addition and reductive elimination are seen in many catalytic cycles such as the Monsanto process and alkene hydrogenation using Wilkinson's catalyst.

Titration

Titration is a common laboratory method of quantitative/ chemical analysis which can be used to determine the concentration of a known reactant. Because volume measurements play a key role in titration, it is also known as *volumetric analysis*. A reagent, called the *titrant*, of known concentration (a standard solution) and volume is used to react with a measured quantity of reactant (the *analyte*).

Using a calibrated burette to add the titrant, it is possible to determine the exact amount that has been consumed when the *endpoint* is reached. The endpoint is the point at which the titration is stopped.

This is classically a point at which the number of moles of titrant is equal to the number of moles of analyte, or some multiple thereof (as in polyprotic acids). In the classic strong acid-strong base titration the endpoint of a titration is when the pH of the reactant is just about equal to 7, and often when the solution permanently changes colour due to an indicator. There are however many different types of titrations.

Many methods can be used to indicate the endpoint of a reaction; titrations often use visual indicators (the reactant mixture changes colour). In simple acid-base titrations a pH indicator may be used, such as phenolphthalein, which becomes pink when a certain pH (about 8.2) is reached or exceeded. Another example is methyl orange, which is red in acids and yellow in alkali solutions.

Not every titration requires an indicator. In some cases, either the reactants or the products are strongly coloured and can serve as the "indicator". For example, an oxidation-reduction titration using potassium permanganate (pink/purple) as the titrant does not require an indicator. When the titrant is reduced, it turns colourless. After the equivalence point, there is excess titrant present. The equivalence point is identified from the first faint pink colour that persists in the solution being titrated.

Due to the logarithmic nature of the pH curve, the transitions are generally extremely sharp, and thus a single drop of titrant just before the *endpoint* can change the pH significantly — leading to an immediate colour change in the indicator. That said, there is a slight difference between the change in indicator colour and the actual equivalence point of the titration. This error is referred to as an indicator error, and it is indeterminate.

History and Etymology

The word "titration" comes from the Latin word "titalus", meaning inscription or title. The French word, titre, also from this origin, means rank. Titration is by definition the determination of rank or concentration of a solution with respect to water with a pH of 7 (which is the pH of pure water), so it corresponds.

The origins of volumetric analysis are in late 18th century French chemistry. Francois Antoine Henri Descroizilles developed the first burette (which looked more like a graduated cylinder) in 1791. Joseph Louis Gay-Lussac developed an improved version of the burette that included a side arm, and coined the terms "pipette" and "burette" in a 1824 paper about on the standarisation of indigo solutions. A major breakthrough in the methodology and popularisation of volumetric analysis was due to Karl Friedrich Mohr, who redesigned the burette by placing a clamp and a tip at the bottom, and wrote the first textbook on the topic, *Lehrbuch der chemisch-analytischen Titrirmethode,* published in 1855.

Preparing a Sample for Titration

In a titration, both titrant and analyte are required to be aqueous, or in a solution form. If the sample isn't a liquid or solution, the samples must be dissolved. If the analyte is very concentrated in the sample, it might be useful to dilute the sample.

Although the vast majority of titrations are carried out in aqueous solution, other solvents such as glacial acetic acid or ethanol (in petrochemistry) are used for special purposes.

A measured amount of the sample can be given in the flask and then be dissolved or diluted. The mathematical result of the titration can be calculated directly with the measured amount. Sometimes the sample is dissolved or diluted beforehand and a measured amount of the solution is used for titration. In this case the dissolving or diluting must be done accurately with a known coefficient because the mathematical result of the titration must be multiplied with this factor.

Many titrations require buffering to maintain a certain pH for the reaction. Therefore, buffer solutions are added to the reactant solution in the flask.

Some titrations require "masking" of a certain ion. This can be necessary when two reactants in the sample would react with the titrant and only one of them must be analysed, or when the reaction would be disturbed or inhibited by this ion. In this case another solution is added to the sample which "masks" the unwanted ion (for instance by a weak binding with it or even forming a solid insoluble substance with it).

Some redox reactions may require heating the solution with the sample and titration while the solution is still hot (to increase the reaction rate).

Procedure

A typical titration begins with a beaker or Erlenmeyer flask containing a precisely known volume of the reactant and a small amount of indicator, placed underneath a burette containing the reagent. By controlling the amount of reagent that is added to the reactant, it is possible to detect the point at which the indicator changes colour. As long as the indicator has been chosen correctly, this should also be the point where the reactant and reagent neutralise each other, and by reading the scale on the burette the volume of reagent can be measured.

As the concentration of the reagent is known, the number of moles of reagent can be calculated (since *concentration = moles / volume*). Then, from the chemical equation involving the two substances, the number of moles present in the reactant can be found. Finally, by dividing the number of moles of reactant by its volume, the concentration is calculated.

Titration Curves

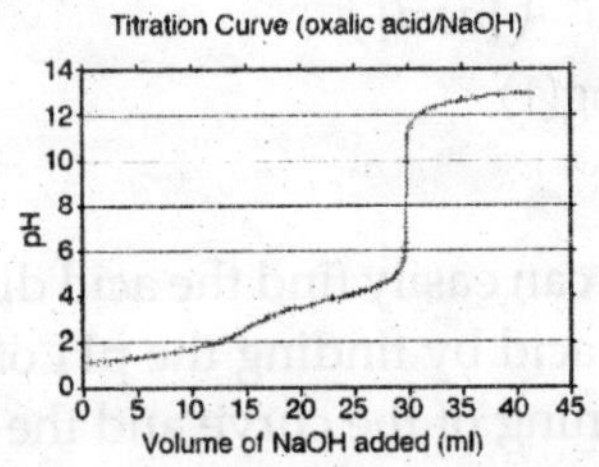

Titrations are often recorded on titration curves, whose compositions are generally identical: the independent variable is the volume of the titrant, while the dependent variable is the pH of the solution (which changes depending on the composition of the two solutions). The equivalence point is a significant point on the graph (the point at which all of the starting solution, usually an acid, has been neutralised by the titrant, usually a base). It can be calculated precisely by finding the second derivative of the titration curve and computing the points of inflection (where the graph changes concavity); however, in most cases, simple visual inspection of the curve will suffice (in the curve given to the right, both equivalence points are visible, after roughly 15 and 30 ml of NaOH solution has been titrated into the oxalic acid solution.) To calculate the pKa values, one must find the volume at the half-equivalence point, that is where half the amount of titrant has been added to form the next compound (here, sodium hydrogen oxalate, then disodium oxalate). Halfway between each equivalence point, at 7.5 ml and 22.5 ml, the pH observed was about 1.5 and 4, giving the pKa values.

In monoprotic acids, the point halfway between the beginning of the curve (before any titrant has been added) and the equivalence point is significant: at that point, the concentrations of the two solutions (the titrant and the original solution) are equal.

Therefore, the Henderson-Hasselbalch equation can be solved in this manner:

$$pH = pK_a + \log\left(\frac{[base]}{[acid]}\right)$$

$$pH = pK_a + \log(1)$$

$$pH = pK_a$$

Therefore, one can easily find the acid dissociation constant of the monoprotic acid by finding the pH of the point halfway between the beginning of the curve and the equivalence point, and solving the simplified equation. In the case of the sample curve, the K_a would be approximately 1.78×10^{-5} from visual inspection (the actual K_{a2} is 1.7×10^{-5})

For polyprotic acids, calculating the acid dissociation constants is only marginally more difficult: the first acid dissociation constant can be calculated the same way as it would be calculated in a monoprotic acid. The second acid dissociation constant, however, is the point halfway between the first equivalence point and the second equivalence point (and so on for acids that release more than two protons, such as phosphoric acid).

Types of Titrations

Titrations can be classified by the type of reaction. Different types of titration reaction include:

- Acid-base titration is based on the neutralisation

reaction between the analyte and an acidic or basic titrant. These most commonly use a pH indicator, a pH metre, or a conductance metre to determine the endpoint.

- A Redox titration is based on an oxidation-reduction reaction between the analyte and titrant. These most commonly use a potentiometer or a redox indicator to determine the endpoint. Frequently either the reactants or the titrant have a colour intense enough that an additional indicator is not needed.
- A Complexometric titration is based on the formation of a complex between the analyte and the titrant. The chelating agent EDTA is very commonly used to titrate metal ions in solution. These titrations generally require specialised indicators that form weaker complexes with the analyte. A common example is Eriochrome Black T for the titration of calcium and magnesium ions.
- A form of titration can also be used to determine the concentration of a virus or bacterium. The original sample is diluted (in some fixed ratio, such as 1:1, 1:2, 1:4, 1:8, etc.) until the last dilution does not give a positive test for the presence of the virus. This value, the titre, may be based on TCID50, EID50, ELD50, LD50 or pfu. This procedure is more commonly known as an assay.

Measuring the Endpoint of a Titration

Different methods to determine the endpoint include:

- *pH Indicator:* This is a substance that changes colour in response to a chemical change. An acid-base indicator (e.g., phenolphthalein) changes colour depending on the pH. Redox indicators are also frequently used. A drop of indicator solution is added to the titration at the start; when the colour changes the endpoint has been reached.
- *A Potentiometer can also be used:* This is an instrument which measures the electrode potential of the solution.

These are used for titrations based on a redox reaction; the potential of the working electrode will suddenly change as the endpoint is reached.

- *pH Metre:* This is a potentiometer which uses an electrode whose potential depends on the amount of H+ ion present in the solution. (This is an example of an ion selective electrode. This allows the pH of the solution to be measured throughout the titration. At the end point there will be a sudden change in the measured pH. It can be more accurate than the indicator method, and is very easily automated.
- *Conductance:* The conductivity of a solution depends on the ions that are present in it. During many titrations, the conductivity changes significantly. (For instance, during an acid-base titration, the H^+ and OH^- ions react to form neutral H_2O. This changes the conductivity of the solution.) The total conductance of the solution depends also on the other ions present in the solution (such as counter ions). Not all ions contribute equally to the conductivity; this also depends on the mobility of each ion and on the total concentration of ions (ionic strength). Thus, predicting the change in conductivity is harder than measuring it.
- *Colour Change:* In some reactions, the solution changes colour without any added indicator. This is often seen in redox titrations, for instance, when the different oxidation states of the product and reactant produce different colours.
- *Precipitation:* If the reaction forms a solid, then a precipitate will form during the titration. A classic example is the reaction between Ag^+ and Cl^- to form the very insoluble salt AgCl. Surprisingly, this usually makes it difficult to determine the endpoint precisely. As a result, precipitation titrations often have to be done as "back" titrations.

- An isothermal titration calorimeter uses the heat produced or consumed by the reaction to determine the endpoint. This is important in biochemical titrations, such as the determination of how substrates bind to enzymes.
- Thermometric titrimetry is an extraordinarily versatile technique. This is differentiated from calorimetric titrimetry by the fact that the heat of the reaction (as indicated by temperature rise or fall) is not used to determine the amount of analyte in the sample solution. Instead, the endpoint is determined by the rate of temperature change.
- Spectroscopy can be used to measure the absorption of light by the solution during the titration, if the spectrum of the reactant, titrant or product is known. The relative amounts of the product and reactant can be used to determine the endpoint.
- Amperometry can be used as a detection technique (amperometric titration). The current due to the oxidation or reduction of either the reactants or products at a working electrode will depend on the concentration of that species in solution. The endpoint can then be detected as a change in the current. This method is most useful when the excess titrant can be reduced, as in the titration of halides with Ag^+. (This is handy also in that it ignores precipitates.)

Other Terms

The term back titration is used when a titration is done "backwards": instead of titrating the original analyte, one adds a known excess of a standard reagent to the solution, then titrates the excess.

A back titration is useful if the end point of the reverse titration is easier to identify than the end point of the normal titration. They are also useful if the reaction between the analyte and the titrant is very slow.

Particular uses

- As applied to biodiesel, titration is the act of determining the acidity of a sample of WVO by the dropwise addition of a known base to the sample while testing with pH paper for the desired neutral pH=7 reading. By knowing how much base neutralises an amount of WVO, we discern how much base to add to the entire batch.
- Titrations in the petrochemical or food industry to define oils, fats or biodiesel and similar substances. An example procedure for all three can be found here:
  - *Acid Number:* An acid-base titration with colour indicator is used to determine the free fatty acid content.
  - *Iodine Number:* A redox titration with colour indication which indicates the amount of unsaturated fatty acids.
  - *Saponification Value:* An acid-base back titration with colour indicator or potentiometric to get a hint about the average chain length of fatty acids in a fat.

Acid-base Titration

An acid-base titration is a method in chemistry that allows quantitative analysis of the concentration of an unknown acid or base solution. It makes use of the neutralisation reaction that occurs between acids and bases, and that we know how acids and bases will react if we know their formula.

Equipment

The key equipment used in a titration are:

- Burette
- White Tile—used to see a colour change in the solution
- Pipette
- Acid/Base Indicator (the one used varies depending on the reactants)

- Erlenmeyer flask
- Standard Solution (a solution of known concentration, a common one is aqueous sodium carbonate (Na_2CO_3)
- Solution of unknown concentration

Method

Before starting the titration a suitable pH indicator must be chosen. The endpoint of the reaction, when all the products have reacted, will have a pH dependent on the relative strengths of the acids and bases. The pH of the endpoint can be roughly determined using the following rules:

- A strong acid reacts with a strong base to form a neutral (pH=7) solution.
- A strong acid reacts with a weak base to form an acidic (pH<7) solution.
- A weak acid reacts with a strong base to form a basic (pH>7) solution.

When a weak acid reacts with a weak base, the endpoint solution will be basic if the base is stronger and acidic if the acid is stronger. If both are of equal strength, then the endpoint pH will be neutral.

A suitable indicator should be chosen, that will experience a change in colour close to the end point of the reaction.

First, the burette should be rinsed with the standard solution, the pipette with the unknown solution, and the conical flask with distilled water.

Secondly, a known volume of the unknown concentration solution should be taken with the pipette and placed into the conical flask, along with a small amount of the indicator chosen. The burette should be filled to the top of its scale with the known solution.

The known solution should then be allowed out of the burette, into the conical flask. At this stage we want a rough estimate of the amount of this solution it took to neutralise the unknown solution. Let the solution out of the burette until the indicator changes colour and then record the value on the burette. This is the first titre and should be discluded from any calculations.

Perform three more titrations, this time more accurately, taking into account we know roughly where the end point will occur. Take note of each of the readings on the burette at the end point, and average these at the end. Endpoint is reached when the indicator just changes colour permanently. This is best achieved by washing a hanging drop from the tip of the burette into the flask right at the end of the titration.

Amperometric Titration

Amperometric titration refers to a class of titrations in which the equivalence point is determined through measurement of the electric current produced by the titration reaction. It is a form of quantitative analysis.

Background

Consider a solution containing the analyte, A, in the presence of some conductive buffer. If an electrolytic potential is applied to the solution through a working electrode, then the measured current depends (in part) on the concentration of the analyte. Measurement of this current can be used to determine the concentration of the analyte directly; this is a form of amperometry. However, the difficulty is that the measured current depends on several other variables, and it is not always possible to control all of them adequately. This limits the precision of direct amperometry.

If the potential applied to the working electrode is sufficient to reduce the analyte, then the concentration of analyte close to the working electrode will decrease. More of the analyte will slowly diffuse into the volume of solution close to the working electrode, restoring the concentration. If the potential applied to the working electrode is great enough (an overpotential), then the concentration of analyte next to the working electrode will depend entirely on the rate of diffusion. In such a case, the current is said to be *diffusion limited*. As the analyte is reduced at the working electrode, the concentration of the analyte in the whole solution will very slowly decrease; this depends on the size of the working electrode compared to the volume of the solution.

What happens if some other species which reacts with the analyte (the titrant) is added? (For instance, chromate ions can be added to oxidise lead ions.) After a small quantity of the titrant (chromate) is added, the concentration of the analyte (lead) has decreased due to the reaction with chromate. The current from the reduction of lead ion at the working electrode will decrease. The addition is repeated, and the current decreases again. A plot of the current against volume of added titrant will be a straight line.

After enough titrant has been added to react completely with the analyte, the excess titrant may itself be reduced at the working electrode. Since this is a different species with different diffusion characteristics (and different half-reaction), the slope of current versus added titrant will have a different slope after the equivalence point. This change in slope marks the equivalence point, in the same way that, for instance, the sudden change in pH marks the equivalence point in an acid-base titration.

The electrode potential may also be chosen such that the titrant is reduced, but the analyte is not. In this case, the presence of excess titrant is easily detected by the increase in current above background (charging) current.

Advantages

The chief advantage over direct amperometry is that the magnitude of the measured current is of interest only as an indicator. Thus, factors that are of critical importance to quantitative amperometry, such as the surface area of the working electrode, completely disappear from amperometric titrations.

The chief advantage over other types of titration is the selectivity offered by the electrode potential, as well as by the choice of titrant.

For instance, lead ion is reduced at a potential of -0.60 V (relative to the saturated calomel electrode), while zinc ions are not; this allows the determination of lead in the presence of zinc. Clearly this advantage depends entirely on the other species present in the sample.

Back Titration

Back titration is an analytical chemistry technique which allows the user to find the concentration of a reactant of unknown concentration by reacting it with an excess volume of another reactant of known concentration. The resulting mixture is then titrated back, taking into account the molarity of the excess which was added.

Back titrations can be used for many reasons, including: when the sample is not soluble in water, when the sample contains impurities that interfere with forward titration, or when the end-point is more easily identified than in forward titration.

Example

Consider using titration to measure the amount of aspirin in a solution. Using titration it would be difficult to identify the end point because aspirin is a weak acid and reactions may proceed slowly.

Using back titration the end-point is more easily recognised in this reaction, as it is a reaction between a strong base and a strong acid. This type of reaction occurs at a high rate and thus produces an end-point which is abrupt and easily seen.

The titration curve for a strong acid with a strong alkali shows that the equivalence point occurs at pH 7. This means that the indicator phenolphthalein can be used. The end-point will be seen when the pink solution produced by the adding of phenolphthalein fades to colourless.

The first stage of this reaction is that of alkaline hydrolysis. This involves reacting the aspirin solution with a measured amount of sodium hydroxide; an amount that will exceed the amount of aspirin present.

Because the hydrolysis reaction occurs at a very low rate at room temperature it will be heated to increase the reaction rate.

$$CH_3COOC_6H_4COOH + 2NaOH \longrightarrow CH_3COONa + HOC_6H_4COONa + H_2O$$

Sodium Ethanoate + Sodium hydroxide ⟶ Sodium acetate + Sodium-2-hydroxy benzoate + Water

The second stage then involves back titration of the hydrolysed sodium hydroxide solution with hydrochloric acid. This process reacts the excess sodium hydroxide with hydrochloric acid.

$$NaOH + HCl \longrightarrow NaCl + H_2O$$

Sodium Hydroxide + Hydrochloric acid ⟶ Sodium Chloride + Water

By the method of back titration the amount of hydrochloric acid needed to neutralise the unreacted sodium hydroxide in the solution can be determined. Knowing this and the amount of sodium hydroxide that was added the amount of aspirin that reacted with the sodium hydroxide can be determined.

Complexometric Titration

Complexometric titration is a type of titration based on complex formation between the analyte and titrant. Complexometric titrations are particularly useful for determination of a mixture of different metal ions in solution. An indicator with a marked colour change is usually used to detect the end-point of the titration.

Any complexation reaction can in theory be applied as a volumetric technique provided that:

1. The reaction reaches equilibrium rapidly following each addition of titrant.
2. Interfering situations do not arise (such as stepwise

formation of various complexes resulting in the presence of more than one complex in solution in significant concentration during the titration process).

3. A complexometric indicator capable of locating equivalence point with fair accuracy is available

In practice, the use of EDTA as a titrant is well established.

Complexometric Titration with EDTA

EDTA, ethylenediamminetetraacetic acid, has four carboxyl groups and two amine groups that can act as electron pair donors, or Lewis bases. The ability of EDTA to potentially donate its six lone pairs of electrons for the formation of coordinate covalent bonds to metal cations makes EDTA a hexadentate ligand.

However, in practice EDTA is usually only partially ionised, and thus forms fewer than six coordinate covalent bonds with metal cations. Disodium EDTA, commonly used in the standardisation of aqueous solutions of transition metal cations, only forms four coordinate covalent bonds to metal cations at pH values less than or equal to 12 as in this range of pH values the amine groups remain protonated and thus unable to donate electrons to the formation of coordinate covalent bonds.

In analytical chemistry the shorthand "Na_2H_2Y" is typically used to designate disodium EDTA. This shorthand can be used to designate any species of EDTA. The "Y" stands for the EDTA molecule, and the "H_2" designates the number of acidic protons bonded to the EDTA molecule.

Bibliography

Aris, Rutherford: *Academic Chemical Engineering in an Historical Perspective*, PDC, New Jersey, 1977.

Aw, S. E.: *Chemical Evolution*, University Education Press, Singapore, 1976.

Ayerst, R. P. Liddell, D. and McLaren M.: *The Role of Chemical Engineering in Providing Propellants and Explosives for the U.K.*, History of Chemical Engineering, Washington, 1980.

Bell, John T. and Scott H. Fogler: *A Virtual Reality Safety and Hazard Analysis Simulation*, American Society for Engineering Education, St. Louis, 2000.

——————— : *Low-Cost Virtual Reality and its Application to Chemical Engineering*, American Institute of Chemical Engineers, New York, 1995.

Bell, John T.: *Introducing Virtual Reality Into the Engineering Curriculum*, Design Manufacturing, Charlottesville, 1996.

Bensaude, V.: *A History of Chemistry*, Harvard University Press, London, 1996.

Beretta, Marco: *The Definition of Chemistry from Agricola to Lavoisier*, Science History Publications, Canton, 1993.

Bird, R. Byron and Lightfoot Edwin N.: *The Role of Transport Phenomena in Chemical Engineering Teaching and Research*, History of Chemical Engineering, Washington, 1980.

Brauman, J. L.: *Frontiers in Chemistry,* Harvard University Press, London, 1988.

Bridgwater, J.: *One Hundred Years of Chemical Engineering,* Kluwer Academic Publishers, Boston, 1989.

Brock, William H.: *The Fontana History of Chemistry,* Fontana Press, London, 1992.

Browne, C.: *A History of the American Chemical Society, Seventy-Five Eventful Years,* American Chemical Society, Washington, 1952.

Bunch, B.: *The Timetables of Technology: A Chronology of the Most Important People and Events in the History of Technology,* Simon and Schuster, New York, 1993.

Calvin, M.: *Chemical Evolution,* Oxford University Press, Oxford, 1969,

Campbell, William A.: *The Chemical Industry,* Longman Group Limited, London, 1971.

Clafin, A. A.: *The General Education Value of the Study of Applied Science, Technology and Industrial Efficiency,* McGraw Hill, New York, 1974.

Coon, A. B. and Stadtherr M. A.: *Generalized Block-tridiagonal Matrix Orderings for Parallel Computation In-process Flowsheeting,* Computers and Chemical Engineering, New York, 1995.

Davies, John T.: *Chemical Engineering: How did it begin and Develop?* A.C.S., Washington, 1980.

Dickerson, R. E. and Geis I.: *Chemistry, Matter and the Universe,* Menlo Park, Benjamin, 1976.

Dickerson, R. E.: *Chemical Evolution and the Origin of Life,* Harvard University Press, London, 1978.

Dodd, Donald B.: *Historical Statistics of the States of the United States,* Greenwood Press, London, 1993.

Donovan, Arthur: *Antoine Lavoisier: Science, Administration and Revolution,* Blackwell, Oxford, 1993.

Du, Pont: *The Autobiography of an American Enterprise,* Charles Scribner's Sons, New York, 1952.

Ewing, A. G. and Wallingford R. A.: *Analytical Chemistry,* B.C.T., Washington, 1989.

Ewing, Galen W.: *Analytical Chemistry,* Chemical and Engineering News, Washington, 1976.

Ferdinando, Abbri: *Negotiating a New Language for Chemistry,* Science History Publications, USA, 1995.

Fox, S. W. Adachi, T. and Stillwell W.: *A Quinone-assisted Photoformation of Energy-rich Chemical Bonds,* Pergamon Press, New York, 1980.

Freshwater, D. C. and George E. Davis: *Norman Swindin, and the Empirical Tradition in Chemical Engineering,* History of Chemical Engineering, ACS, Washington, 1980.

Freshwater, D. C.: *The Development of Chemical Engineering as Shown by its Texts,* Kluwer Academic Publishers, Boston, 1989.

Fuller, E. C.: *Chemistry and Man's Environment,* Houghton Mifflin, Boston, 1974.

Furter, William F.: *Chemical Engineering and the Public Image,* History of Chemical Engineering, Washington, 1980.

Futrell, A. W.: *Orientation to Engineering,* Charles E. Merrill Books, Columbus, 1961.

Glajch, J. L. and Kirkland J. J.: *Analytical Chemistry,* Pergamon Press, New York, 1983.

Goldberg, E. D.: *Chemistry in the Oceans,* American Association for the Advancement of Science, Washington, 1961.

Gornowski, Edward J.: *The History of Chemical Engineering at Exxon,* History of Chemical Engineering, Washington, 1980.

Guedon, Jean-Claude: *Conceptual and Institutional Obstacles to the Emergence of Unit Operations in Europe,* History of Chemical Engineering, Washington,1980.

Haber, L. F.: *The Chemical Industry International Growth and Technological Change,* Claredon Press, Oxford, 1971.

Hammond, John H.: *The Chemical Engineer,* John Wiley & Sons, New York, 1929.

Haugland, R. P.: *Handbook of Fluorescent Probes and Research Chemicals,* Molecular Probes, Eugene, 1985.

Hayes, Williams: *American Chemical Industry,* D. Van Nostrand Company, New York, 1954.

Herrnstein, Richard J.: *The Bell Curve: Intelligence and Class Structure in American Life,* The Free Press, New York, 1994.

Hiscox, Gardner D: *Henley's Formulas for Home and Workshop,* Wings Books, New York, 1907.

Holmes, Frederic Lawrence: *The Next Crucial Year, the Sources of his Quantitative Method in Chemistry,* Princeton University Press, Princeton, 1998.

Hougen, Olaf A.: *Seven Decades of Chemical Engineering,* Chemical Engineering Progress, Washington, 1972.

Hu, Y. F. Maguire K. C. F. and R. J. Blake: *A Multilevel Unsymmetric Matrix Ordering Algorithm for Parallel Process Simulation,* Pergamon Press, New York, 1999.

Jefferson, Edward G.: *The Emergence of Chemical Engineering as a Multidiscipline,* Chemical Engineering Progress, London, 1988.

Johnson, William S.: *The Pollution of Streams by Manufacturing Wastes,* McGraw Hill, New York, 1911.

Kernighan, B. W.: *An Efficient Heuristic Procedure for Partitioning Graphs, Bell Systems,* Pergamon Press, New York, 1970.

Knight, David M.: *Ideas in Chemistry: A History of the Science,* Rutgers University Press, New Brunswick, 1992.

Knox, J. H. and Kauer B.: *High Performance Liquid Chromatography,* Wiley Interscience, New York, 1989.

Krieger, James H.: *Report on Chemical Engineering Reflects a Profession in Flux,* Pergamon Press, New York, 1987.

Landau, Ralph: *The Chemical Engineer: Today and Tomorrow*, Chemical Engineering Progress, London, 1972.

Layton, Edwin T.: *Technology and Civilization: Renaissance to Industrial Revolution*, University of Minnesota, Minnesota, 1994.

—————: *The Dimensional Revolution: The New Relations Between Theory and Experiment in Engineering in the Age of Michelson*, Pergamon Press, New York, 1988.

Levenspiel, Octave: *The Coming of Age of Chemical Reaction Engineering*, Chemical Engineering Science, 1980.

Levere, Trevor H.: *Transforming Matter: A History of Chemistry from Alchemy to the Buckyball*, Johns Hopkins University Press, Baltimore, 2001.

Lewis, W.: *An Object Lesson in Efficiency, Technology and Industrial Efficiency*, McGraw Hill, New York, 1911.

Lyon, Tracy.: *Scientific Industrial Operation*, McGraw Hill, New York, 1911.

Lyons, John W.: *Fire*, Scientific American, New York, 1985.

Maclaurin, Richard C.: *Some Factors in the Institute's Success*, McGraw Hill, New York, 1911.

Mallya, J. U. and Stadtherr M. A.: *A Multifrontal Approach for Simulating Equilibrium-stage Processes on Supercomputers*, Industrial and Engineering Chemistry Research, 1997.

Mant, C. T. and Hodges R. S.: *High-Performance Liquid Chromatography of Peptides and Proteins*, CRC Press, Boston, 1991.

Mark, Herman F.: *Polymer Chemistry*, Chemical and Engineering News, 1976.

Morrison, A. Cressy: *Man in a Chemical World*, Charles Scribner's Sons, New York, 1937.

Munroe, J. P.: *Influence of the Institute upon the Development of Modern Education*, McGraw Hill, New York, 1911.

Nye, Mary Jo.: *The Pursuit of Modern Chemistry and Physics*, Twayne, New York, 1996.

Othmer, Donald F.: *The Big Future Program for Chemical Engineers: Fuel and Energy Conversions*, History of Chemical Engineering, Washington, 1980.

Peppas, Nicholas A. and Harland R. S.: *Unit Process Against Unit Operations: The Educational Fights of the Thirties*, Kluwer Academic Publishers, Boston, 1989.

Peppas, Nicholas A.: *Academic Connections of the 20th Century U.S. Chemical Engineers*, Kluwer Academic Publishers, Boston, 1989.

———————: *The Origins of Academic Chemical Engineering*, Kluwer Academic Publishers, Boston, 1989.

Perry, Robert H. and Green, Don: *Perry's Chemical Engineers' Handbook*, McGraw Hill, New York, 1984.

Person, James E.: *Statistical Forcasts of the United States*, Gale Research, Detroit, 1993.

Pigford, Robert L.: *Chemical Technology*, Chemical and Engineering News, 1976.

Principe, Lawrence M.: *The Aspiring Adept: Robert Boyle and His Alchemical Quest*, Princeton University Press, Princeton, 1998.

Rae, H. K. *Three Decades of Canadian Nuclear Chemical Engineering*, History of Chemical Engineering, Washington, 1980.

Raloff, J.: *Is There a Cosmic Chemistry of Life?*, Science News,1986.

Reynolds, Terry S. *Defining Professional Boundaries: Chemical Engineering in the Early 20th Century*, Technology and Culture, 1986.

———————: *A History of the American Institute of Chemical Engineers*, American Institute of Chemical Engineers, New York, 1983.

Richards, E. H.: *The Elevation of Applied Science to an Equal Rank with the So-called Learned Professions*, McGraw Hill, New York, 1911.

Rocke, Alan J.: *Chemical Atomism in the Nineteenth Century*, Ohio State University Press, Columbus, 1984.

Schmidt, Lanny D.: *The Engineering of Chemical Reactions*, University of Minnesota, 1995.

Schoeff, M. S. and Williams, R. H.: *Principles of Laboratory Instruments*, St. Louis, Mosby, 1993.

Schoenemann, Karl: The *Separate Development of Chemical Engineering in Germany*, History of Chemical Engineering, Washington, 1980.

Schramm, D. N.: *The Age of the Elements*, Scientific American, 1974.

Scott, Fogler H. and Bell, John T.: *Preliminary Testing of a Virtual Reality based Module for Safety and Hazard Evaluation*, Bradley University, Peoria, 1996.

—————: *Virtual Reality in Chemical Engineering Education*, Purdue University, Indiana, 1995.

Scott, Fogler H.: *The Application of Virtual Reality to Chemical Engineering and Education*, Unpublished, Miami, 1998.

—————: *Virtual Laboratory Accidents Designed to Increase Safety Awareness*, American Society for Engineering Education, Charlotte, 1999.

—————: *Virtual Reality in Chemical Engineering Education*, University of Detroit, Mercy, 1998.

Senecal, Vance E.: *Du Pont and Chemical Engineering in the Twentieth Century*, History of Chemical Engineering, Washington, 1980.

Servos, John W.: *The Industrial Relations of Scienc*, M.I.T, Washington, 1980.

Skinner, H. J.: *The Debt of the Manufacturer to the Chemist*, McGraw Hill, New York, 1911.

Smith, H. E. *The Chemist in the Service of the Railroad*, McGraw Hill, New York, 1911.

Talbot, H. P. *The Engineering School Graduate: His Strength and His Weakness,* McGraw Hill, New York, 1911.

Tarbell, D. Stanley: *Organic Chemistry,* Chemical and Engineering News, 1976.

Van Antwerpen, F. J.: *The Origins of Chemical Engineering, History of Chemical Engineering,* Washington, 1980.

Van, Spronsen, J. W.: *The Periodic System of the Chemical Elements,* Elsevier, New York, 1969.

Waldrop, M. M., *Catalytic RNA Wins Chemistry Nobel,* Science, 1989.

Walker, W. H.: *The Spirit of Alchemy in Modern Industry,* McGraw Hill, New York, 1911.

Weber, H. C: *The Improbable Achievement,* M.I.T., Washington, 1980.

Westwater, J. W.: *The Beginnings of Chemical Engineering Education in the USA,* History of Chemical Engineering, Washington, 1980.

White, Robert M.: *Technological Competitiveness and Chemical Engineering,* Chemical Engineering Progress, Washington, 1988.

Willard, H. Merritt L. Dean, J.A. and Settle F.A.: *Instrumental Methods of Analysis,* Wadsworth Publishing, 1988.

Williams, Glenn C. and Vivian, J. Edward: *Pioneers in Chemical Engineering,* History of Chemical Engineering, Washington, 1980.

Zitney, S. E., Choudhary S. and Stadtherr M. A.: *A Parallel Frontal Solver for Large-scale Process Simulation and Optimization,* Aiche Journal, 1997.

Zlokarnik, Marko: *Dimensional Analysis and Scale-up in Chemical Engineering,* Springer Verlag, New York, 1991.

Zubieta, Jon A. amd Juckerman, Jerold J.: *Inorganic Chemistry,* Chemical and Engineering News, 1976.

Index

A

B

C

D

E

F

G

H

Bibliography

Aris, Rutherford: *Academic Chemical Engineering in an Historical Perspective*, PDC, New Jersey, 1977.

Aw, S. E.: *Chemical Evolution*, University Education Press, Singapore, 1976.

Ayerst, R. P. Liddell, D. and McLaren M.: *The Role of Chemical Engineering in Providing Propellants and Explosives for the U.K.*, History of Chemical Engineering, Washington, 1980.

Bell, John T. and Scott H. Fogler: *A Virtual Reality Safety and Hazard Analysis Simulation*, American Society for Engineering Education, St. Louis, 2000.

———: *Low-Cost Virtual Reality and its Application to Chemical Engineering*, American Institute of Chemical Engineers, New York, 1995.

Bell, John T.: *Introducing Virtual Reality Into the Engineering Curriculum*, Design Manufacturing, Charlottesville, 1996.

Bensaude, V.: *A History of Chemistry*, Harvard University Press, London, 1996.

Beretta, Marco: *The Definition of Chemistry from Agricola to Lavoisier*, Science History Publications, Canton, 1993.

Bird, R. Byron and Lightfoot Edwin N.: *The Role of Transport Phenomena in Chemical Engineering Teaching and Research*, History of Chemical Engineering, Washington, 1980.

Brauman, J. L.: *Frontiers in Chemistry,* Harvard University Press, London, 1988.

Bridgwater, J.: *One Hundred Years of Chemical Engineering,* Kluwer Academic Publishers, Boston, 1989.

Brock, William H.: *The Fontana History of Chemistry,* Fontana Press, London, 1992.

Browne, C.: *A History of the American Chemical Society, Seventy-Five Eventful Years,* American Chemical Society, Washington, 1952.

Bunch, B.: *The Timetables of Technology: A Chronology of the Most Important People and Events in the History of Technology,* Simon and Schuster, New York, 1993.

Calvin, M.: *Chemical Evolution,* Oxford University Press, Oxford, 1969,

Campbell, William A.: *The Chemical Industry,* Longman Group Limited, London, 1971.

Clafin, A. A.: *The General Education Value of the Study of Applied Science, Technology and Industrial Efficiency,* McGraw Hill, New York, 1974.

Coon, A. B. and Stadtherr M. A.: *Generalized Block-tridiagonal Matrix Orderings for Parallel Computation In-process Flowsheeting,* Computers and Chemical Engineering, New York, 1995.

Davies, John T.: *Chemical Engineering: How did it begin and Develop?* A.C.S., Washington, 1980.

Dickerson, R. E. and Geis I.: *Chemistry, Matter and the Universe,* Menlo Park, Benjamin, 1976.

Dickerson, R. E.: *Chemical Evolution and the Origin of Life,* Harvard University Press, London, 1978.

Dodd, Donald B.: *Historical Statistics of the States of the United States,* Greenwood Press, London, 1993.

Donovan, Arthur: *Antoine Lavoisier: Science, Administration and Revolution,* Blackwell, Oxford, 1993.

Du, Pont: *The Autobiography of an American Enterprise*, Charles Scribner's Sons, New York, 1952.

Ewing, A. G. and Wallingford R. A.: *Analytical Chemistry*, B.C.T., Washington, 1989.

Ewing, Galen W.: *Analytical Chemistry*, Chemical and Engineering News, Washington, 1976.

Ferdinando, Abbri: *Negotiating a New Language for Chemistry*, Science History Publications, USA, 1995.

Fox, S. W. Adachi, T. and Stillwell W.: *A Quinone-assisted Photoformation of Energy-rich Chemical Bonds*, Pergamon Press, New York, 1980.

Freshwater, D. C. and George E. Davis: *Norman Swindin, and the Empirical Tradition in Chemical Engineering*, History of Chemical Engineering, ACS, Washington, 1980.

Freshwater, D. C.: *The Development of Chemical Engineering as Shown by its Texts*, Kluwer Academic Publishers, Boston, 1989.

Fuller, E. C.: *Chemistry and Man's Environment*, Houghton Mifflin, Boston, 1974.

Furter, William F.: *Chemical Engineering and the Public Image*, History of Chemical Engineering, Washington, 1980.

Futrell, A. W.: *Orientation to Engineering*, Charles E. Merrill Books, Columbus, 1961.

Glajch, J. L. and Kirkland J. J.: *Analytical Chemistry*, Pergamon Press, New York, 1983.

Goldberg, E. D.: *Chemistry in the Oceans*, American Association for the Advancement of Science, Washington, 1961.

Gornowski, Edward J.: *The History of Chemical Engineering at Exxon*, History of Chemical Engineering, Washington, 1980.

Guedon, Jean-Claude: *Conceptual and Institutional Obstacles to the Emergence of Unit Operations in Europe*, History of Chemical Engineering, Washington,1980.

Haber, L. F.: *The Chemical Industry International Growth and Technological Change,* Claredon Press, Oxford, 1971.

Hammond, John H.: *The Chemical Engineer,* John Wiley & Sons, New York, 1929.

Haugland, R. P.: *Handbook of Fluorescent Probes and Research Chemicals,* Molecular Probes, Eugene, 1985.

Hayes, Williams: *American Chemical Industry,* D. Van Nostrand Company, New York, 1954.

Herrnstein, Richard J.: *The Bell Curve: Intelligence and Class Structure in American Life,* The Free Press, New York, 1994.

Hiscox, Gardner D: *Henley's Formulas for Home and Workshop,* Wings Books, New York, 1907.

Holmes, Frederic Lawrence: *The Next Crucial Year, the Sources of his Quantitative Method in Chemistry,* Princeton University Press, Princeton, 1998.

Hougen, Olaf A.: *Seven Decades of Chemical Engineering,* Chemical Engineering Progress, Washington, 1972.

Hu, Y. F. Maguire K. C. F. and R. J. Blake: *A Multilevel Unsymmetric Matrix Ordering Algorithm for Parallel Process Simulation,* Pergamon Press, New York, 1999.

Jefferson, Edward G.: *The Emergence of Chemical Engineering as a Multidiscipline,* Chemical Engineering Progress, London, 1988.

Johnson, William S.: *The Pollution of Streams by Manufacturing Wastes,* McGraw Hill, New York, 1911.

Kernighan, B. W.: *An Efficient Heuristic Procedure for Partitioning Graphs, Bell Systems,* Pergamon Press, New York, 1970.

Knight, David M.: *Ideas in Chemistry: A History of the Science,* Rutgers University Press, New Brunswick, 1992.

Knox, J. H. and Kauer B.: *High Performance Liquid Chromatography,* Wiley Interscience, New York, 1989.

Krieger, James H.: *Report on Chemical Engineering Reflects a Profession in Flux,* Pergamon Press, New York, 1987.

Landau, Ralph: *The Chemical Engineer: Today and Tomorrow,* Chemical Engineering Progress, London, 1972.

Layton, Edwin T.: *Technology and Civilization: Renaissance to Industrial Revolution,* University of Minnesota, Minnesota, 1994.

————————: *The Dimensional Revolution: The New Relations Between Theory and Experiment in Engineering in the Age of Michelson,* Pergamon Press, New York, 1988.

Levenspiel, Octave: *The Coming of Age of Chemical Reaction Engineering,* Chemical Engineering Science, 1980.

Levere, Trevor H.: *Transforming Matter: A History of Chemistry from Alchemy to the Buckyball,* Johns Hopkins University Press, Baltimore, 2001.

Lewis, W.: *An Object Lesson in Efficiency, Technology and Industrial Efficiency,* McGraw Hill, New York, 1911.

Lyon, Tracy.: *Scientific Industrial Operation,* McGraw Hill, New York, 1911.

Lyons, John W.: *Fire,* Scientific American, New York, 1985.

Maclaurin, Richard C.: *Some Factors in the Institute's Success,* McGraw Hill, New York, 1911.

Mallya, J. U. and Stadtherr M. A.: *A Multifrontal Approach for Simulating Equilibrium-stage Processes on Supercomputers,* Industrial and Engineering Chemistry Research, 1997.

Mant, C. T. and Hodges R. S.: *High-Performance Liquid Chromatography of Peptides and Proteins,* CRC Press, Boston, 1991.

Mark, Herman F.: *Polymer Chemistry,* Chemical and Engineering News, 1976.

Morrison, A. Cressy: *Man in a Chemical World,* Charles Scribner's Sons, New York, 1937.

Munroe, J. P.: *Influence of the Institute upon the Development of Modern Education,* McGraw Hill, New York, 1911.

Nye, Mary Jo.: *The Pursuit of Modern Chemistry and Physics,* Twayne, New York, 1996.

Othmer, Donald F.: *The Big Future Program for Chemical Engineers: Fuel and Energy Conversions,* History of Chemical Engineering, Washington, 1980.

Peppas, Nicholas A. and Harland R. S.: *Unit Process Against Unit Operations: The Educational Fights of the Thirties,* Kluwer Academic Publishers, Boston, 1989.

Peppas, Nicholas A.: *Academic Connections of the 20th Century U.S. Chemical Engineers,* Kluwer Academic Publishers, Boston, 1989.

———————: *The Origins of Academic Chemical Engineering,* Kluwer Academic Publishers, Boston, 1989.

Perry, Robert H. and Green, Don: *Perry's Chemical Engineers' Handbook,* McGraw Hill, New York, 1984.

Person, James E.: *Statistical Forcasts of the United States,* Gale Research, Detroit, 1993.

Pigford, Robert L.: *Chemical Technology,* Chemical and Engineering News, 1976.

Principe, Lawrence M.: *The Aspiring Adept: Robert Boyle and His Alchemical Quest,* Princeton University Press, Princeton, 1998.

Rae, H. K. *Three Decades of Canadian Nuclear Chemical Engineering,* History of Chemical Engineering, Washington, 1980.

Raloff, J.: *Is There a Cosmic Chemistry of Life?,* Science News,1986.

Reynolds, Terry S. *Defining Professional Boundaries: Chemical Engineering in the Early 20th Century,* Technology and Culture, 1986.

———————: *A History of the American Institute of Chemical Engineers,* American Institute of Chemical Engineers, New York, 1983.

Richards, E. H.: *The Elevation of Applied Science to an Equal Rank with the So-called Learned Professions,* McGraw Hill, New York, 1911.

Rocke, Alan J.: *Chemical Atomism in the Nineteenth Century,* Ohio State University Press, Columbus, 1984.

Schmidt, Lanny D.: *The Engineering of Chemical Reactions,* University of Minnesota, 1995.

Schoeff, M. S. and Williams, R. H.: *Principles of Laboratory Instruments,* St. Louis, Mosby, 1993.

Schoenemann, Karl: The *Separate Development of Chemical Engineering in Germany,* History of Chemical Engineering, Washington, 1980.

Schramm, D. N.: *The Age of the Elements,* Scientific American, 1974.

Scott, Fogler H. and Bell, John T.: *Preliminary Testing of a Virtual Reality based Module for Safety and Hazard Evaluation,* Bradley University, Peoria, 1996.

———————— : *Virtual Reality in Chemical Engineering Education,* Purdue University, Indiana, 1995.

Scott, Fogler H.: *The Application of Virtual Reality to Chemical Engineering and Education,* Unpublished, Miami, 1998.

————————: *Virtual Laboratory Accidents Designed to Increase Safety Awareness,* American Society for Engineering Education, Charlotte, 1999.

———————— : *Virtual Reality in Chemical Engineering Education,* University of Detroit, Mercy, 1998.

Senecal, Vance E.: *Du Pont and Chemical Engineering in the Twentieth Century,* History of Chemical Engineering, Washington, 1980.

Servos, John W.: *The Industrial Relations of Scienc,* M.I.T, Washington, 1980.

Skinner, H. J.: *The Debt of the Manufacturer to the Chemist,* McGraw Hill, New York, 1911.

Smith, H. E. *The Chemist in the Service of the Railroad,* McGraw Hill, New York, 1911.

Talbot, H. P. *The Engineering School Graduate: His Strength and His Weakness*, McGraw Hill, New York, 1911.

Tarbell, D. Stanley: *Organic Chemistry*, Chemical and Engineering News, 1976.

Van Antwerpen, F. J.: *The Origins of Chemical Engineering, History of Chemical Engineering*, Washington, 1980.

Van, Spronsen, J. W.: *The Periodic System of the Chemical Elements*, Elsevier, New York, 1969.

Waldrop, M. M., *Catalytic RNA Wins Chemistry Nobel*, Science, 1989.

Walker, W. H.: *The Spirit of Alchemy in Modern Industry*, McGraw Hill, New York, 1911.

Weber, H. C: *The Improbable Achievement*, M.I.T., Washington, 1980.

Westwater, J. W.: *The Beginnings of Chemical Engineering Education in the USA*, History of Chemical Engineering, Washington, 1980.

White, Robert M.: *Technological Competitiveness and Chemical Engineering*, Chemical Engineering Progress, Washington, 1988.

Willard, H. Merritt L. Dean, J.A. and Settle F.A.: *Instrumental Methods of Analysis*, Wadsworth Publishing, 1988.

Williams, Glenn C. and Vivian, J. Edward: *Pioneers in Chemical Engineering*, History of Chemical Engineering, Washington, 1980.

Zitney, S. E., Choudhary S. and Stadtherr M. A.: *A Parallel Frontal Solver for Large-scale Process Simulation and Optimization*, Aiche Journal, 1997.

Zlokarnik, Marko: *Dimensional Analysis and Scale-up in Chemical Engineering*, Springer Verlag, New York, 1991.

Zubieta, Jon A. amd Juckerman, Jerold J.: *Inorganic Chemistry*, Chemical and Engineering News, 1976.

Index

C

D

E

I

K

L

M

N

O

P

Q

R

S

T

V

W

Y

Z

□□□